Heat-Treatment of Steel

A Comprehensive Treatise on the Hardening, Tempering, Annealing and Casehardening of Various Kinds of Steel, Including High-Speed, High-Carbon, Alloy and Low-Carbon Steels, together with Chapters on Heat-Treating Furnaces and on Hardness Testing

By Erik Oberg

Published by Pantianos Classics

ISBN-13: 978-1-78987-556-0

First published in 1914

Contents

Preface

In the development that has taken place in the methods and processes pertaining to the machine building trades during the past fifteen or twenty years, most remarkable changes have been wrought in the heat-treatment of steel, including the hardening, tempering, annealing and casehardening of the various kinds of steels. The introduction of high-speed steel and of the various alloy steels has especially demanded great modifications of past practice. The present book places on record the modern methods now employed in the heat-treatment of steel, and includes also a treatise on the methods used for measuring the hardness of metals by the various hardness testing apparatus that have been developed in this country and abroad.

Special attention has been given to a number of methods very recently developed, making this book the most modern and complete on the subject; thus, for example, a very comprehensive treatment is given of electric hardening furnaces, a development unknown only a few years ago. Another of the more recent developments to which attention has been given is the method of casehardening by carbonaceous gas which has been developed very recently.

The well-known twenty-five cent Reference Books which Machinery has published since 1908 and of which one hundred and twenty-five different titles have been published during the past six years, include the best of the material that has appeared in Machinery in past years, adequately revised, amplified and brought up-to-date. Many subjects, however, cannot be covered to an adequate extent in all their phases in books of this size, and in answer to a demand for more comprehensive and detailed treatments on the more important mechanical subjects, it has been deemed advisable to bring out a number of larger volumes, each covering one subject completely. This book is one of these volumes.

The information contained in this book is mainly compiled from articles published in Machinery and the best on the subject that has appeared in the Reference Books is also included. Amplifications and additions have been made wherever necessary. For the material contained, Machinery is indebted to a large number of men who have furnished information to its columns. In many cases it has not been possible to give credit to each individual contributor, but it should be mentioned that some of the most important chapters have been mainly compiled from articles by Ralph Badger and E. F. Lake. To all other writers whose material has appeared in Machinery and is now used in this book, the publishers hereby express their appreciation.

Machinery

New York, June, 1914

Chapter One - Hardening Carbon Steels

Originally the name steel was applied to various combinations of iron and carbon, there being present, together with these, as impurities, small proportions of silicon and manganese. At the present time, however, the use of the name is extended to cover combinations of iron with tungsten, vanadium, nickel, chromium, molybdenum, titanium and some of the rarer elements. These latter combinations are quite generally known as the *alloy* steels to distinguish them from the *carbon* steels, in which latter the characteristic properties are dependent upon the presence of carbon alone. The alloy steels are divided into high-speed steels and low-carbon alloy steels. The specific properties that distinguish these different steels are due in part to their respective compositions, that is, to the particular elements they contain, and, in part, to their subsequent working and heat-treatment.

Effect of Difference in Composition of Steel. — In general, any change in the composition of a steel results in some change in its properties. For example, the addition of certain metallic elements to a carbon steel causes, in the alloy steel thus formed, a change in position of the proper hardening temperature point. Tungsten or manganese tend to lower this point, boron and vanadium to raise it; the amount of the change is, generally, proportional to the amount of the element added. Just as a small proportion of carbon added to iron produces steel which has decidedly different properties from those found in pure iron, so increasing the proportion of carbon in the steel thus formed, within certain limits, causes a variation in the degree in which these properties manifest themselves. For example, consider the property of tensile strength. In a "ten-point" carbon steel (one in which there is present but 0.10 per cent of carbon), the tensile strength is very nearly 25 per cent greater than in pure iron. Adding more carbon causes the tensile strength to rise, approximately, at the rate of 2.5 per cent for each 0.01 per cent of carbon added.

Carbon steels are divided into three classes according to the proportion of carbon which they contain. The first of these embraces the "unsaturated" steels, in which the carbon content is lower than 0.89 per cent; the second, the "saturated" steels, in which the proportion of carbon is exactly 0.89 per cent; and, the third, the "supersaturated" steels, in which the carbon content is higher than 0.89 per cent.

Effect of Heat-treatment. — With a steel of a given composition, proper heat-treatments may be applied which, of themselves, will first alter in form or degree some of its specific properties, or second, practically eliminate one or more of these, or third, add certain new ones. Physical properties of size, shape and ductility are examples of the first case; an example of the second case is found in the heating of steel beyond its hardening temperature, which takes away its magnetism, making it non-magnetic; and an example of the third case is the fact that a greater degree of hardness may be added to steel by the process of hardening. In this connection it must be understood that, strictly speaking, hardness is a relative term and all steel has some hardness.

There are three general heat-treatment operations: annealing, hardening — with which this chapter will deal — and tempering. In all of these the object sought is to change in some manner the existing properties of the steel; in other words, to produce in it certain permanent conditions.

The controlling factor in all heat-treatment is temperature. Whether the operation is annealing, hardening or tempering, there is for any certain steel and particular use thereof a definite temperature point that alone gives the best results. Insufficient temperatures do not produce the results sought. Excessive temperatures, either through ignorance of what the correct point is or through inability to tell when it exists, cause "burned" steel; this is a common failing, resulting in great loss. Very slight variations from the proper temperature may do irreparable damage.

Due to temperature variation alone, carbon steel may be had in any of three conditions: first, in the unhardened or annealed state, when not heated to temperatures above 1350 degrees F.; second, in the hardened state, by heating to temperatures between 1350 and 1500 degrees F.; third, in a state softer than the second though harder than the first, when heated to temperatures which exceed 1500 degrees F.

The Hardening Process. — The hardening of a carbon steel is the result of a change of internal structure which takes place in the steel when heated properly to a correct temperature. In the different carbon steels this change, for practical purposes, is effective only in those in which the proportion of carbon is between 0.20 per cent and 2.0 per cent, that is, between "twenty-point" and "two" carbon steels, respectively.

When heated, ordinary carbon steels begin to soften at about 390 degrees F. and continue to soften throughout a range of 310 degrees F. At the point 700 degrees F. practically all of the hardness has disappeared. "Red hardness" in a steel is a property which enables it to remain hard at red heat. In a high-speed steel this property is of the first importance, 1020 degrees F. being a minimum temperature at which softening may begin. This is some 630 degrees F. above the point at which softening commences in ordinary carbon steels.

The process of hardening steel consists essentially of heating the steel to the required temperature and quenching it suddenly in some cooling medium. The methods of heating and the different kinds of quenching baths used will be explained in detail later. Generally speaking, the furnaces used for the heating of steel for hardening are heated either by gas, oil, electricity or solid fuel. Each of these methods has its advantages, according to the local conditions, the requirements on the work, the quantities of tools to be hardened, the cost of fuel, etc. Electricity offers many attractive advantages for the heating of steel. The electric resistance furnace, as now built in a variety of sizes of either muffle or two-chamber types, has one fundamental advantage over coal, coke, gas or oil-heated furnaces, which by many is claimed to render it quite superior. It is entirely free from all products of combustion, the heat being produced by electrical resistance. This is important, as it does away with the chief cause of oxidation of the heated steel. Further, the temperature of the electric furnaces can be easily and accurately regulated to, and maintained uniform at, any desired point. When electric power is generated for other purposes, the increased cost of this form of energy for operating furnaces is not sufficient to argue against it. Even when the current is purchased, the superior quality of work performed by this kind of furnace is claimed to frequently more than offset the slightly higher cost of operation.

In the actual heating of a piece of steel, several requirements are essential to good hardening: first, that small projections or cutting edges are not heated more rapidly than is the body of the piece, that is, that all parts are heated at the same rate, and second, that all parts are heated to the same temperature. These conditions are facilitated by slow heating, especially when the heated piece is large. A uniform heat, as low in temperature as will give the required hardness, produces the best product. Lack of uniformity in heating causes irregular grain and internal strains, and may even produce surface cracks. Any temperature above the "critical point" of steel tends to open its grain — to make it coarse and to diminish its strength — though such a temperature may not be sufficient to lessen appreciably its hardness.

Critical Temperatures. — The temperatures at which take place the previously mentioned internal changes in the structure of a steel are frequently spoken of as the "critical" points. These are different in steels of different carbon contents. The higher the percentage of carbon present, the lower the temperature required to produce the internal change. In other words, the critical points of a high-carbon steel are lower than those of a low-carbon steel. In steels of the commonly used carbon contents, there are two of these critical temperatures, called the *decalescence* point and the *recalescence* point, respectively.

Decalescence and Its Relation to Hardening. — Everyone interested in the hardening of steel will have noticed the increasing frequency with which reference is made to the decalescence and recalescence points of steel, in articles appearing in the technical press from time to time. It is only during the past few years that this peculiarity in steel has come to the front, and there are still very many who do not possess even a rudimentary knowledge of the subject. The somewhat obscure references one usually sees in the treatises on hardening will not help the man in the hardening shop very much to a better understanding of the matter, and therefore an elementary explanation of the phenomenon will be welcome to many. It may be quoted that, as a matter of history, hardening has been done with more or less success, from the days of the famous Damascus swords up to only a comparatively short time ago, without anyone having discovered that steel possessed such a peculiarity as decalescence, but nevertheless its relation to hardening has always existed, and its discovery paved the way for much scientific investigation into a subject that had been previously controlled by rule of thumb.

The "decalescence" and "recalescence" or "critical" points (also sometimes designated Ac. 1 and Ar. 1), that bear relation to the hardening of steel, are simply evolutions that occur in the chemical composition of steel at certain temperatures during both heating and cooling. Steel at normal temperatures carries its carbon, which is its chief hardening component, in a certain form — pearlite carbon to be more explicit — and if heated to a certain temperature, a change occurs and the pearlite carbon becomes cementite or hardening carbon. Likewise, if allowed to cool slowly, the hardening carbon changes back again to pearlite. The points at which these evolutions occur are the decalescence and recalescence or critical points, and the effect of these molecular changes is to cause an increased absorption of heat on a rising temperature and an evolution of heat on a falling temperature. That is to say, during the heating of a piece of steel a halt occurs, and it continues to absorb heat without appreciably rising in temperature, at the decalescence point, although its immediate surroundings may be hotter than the steel. Likewise, steel cooling slowly will, at a certain temperature, actually increase in temperature although its surroundings may be colder. This takes place at the recalescence point.

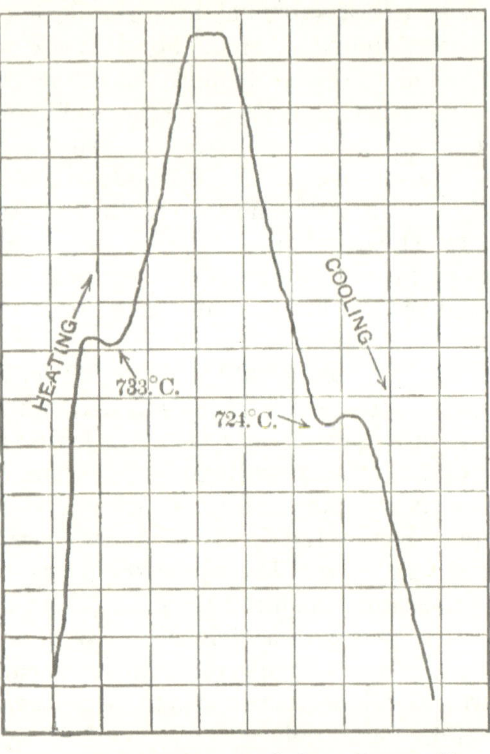

Fig. 1. Curve made by a Recording Pyrometer showing the Decalescence and Recalescence Points

In Fig. 1 is shown a curve, taken on a recording pyrometer, in which the decalescence and recalescence points are well developed. From this it will be seen that the absorption of heat occurred at a point marked 733 degrees C. (1351 degrees F.) on the rising temperature, and the evolution of heat at 724 degrees C. (1335 degrees F.) on the falling temperature. The relation of these critical points to hardening is in the fact that unless a temperature sufficient to produce the first action is reached, so that the pearlite carbon will be changed to hardening carbon, and unless it is cooled with sufficient rapidity to practically eliminate the second action, no hardening can take place. The rate of cooling is material and accounts for the fact that large articles require to be quenched at higher temperatures than small ones.

A very important feature is that steel containing hardening carbon, *i.e.*, steel above the temperature of decalescence, is non-. magnetic. Anyone may demonstrate this for himself by heating a piece of steel to a bright red and testing it with an ordinary magnet. While bright red it will be found to have no attraction for the magnet, but at about a cherry-red it regains its magnetic properties. This feature has been taken advantage of as a means of determining the correct hardening temperature, and appliances for its application are on the market. Its use is certainly to be recommended where no installation of pyrometers exists; the only point requiring judgment is the length of time an article should remain in the furnace after it has become non-magnetic. This varies with the weight and cooling surface, but may be tabulated according to weight, leaving very little to personal judgment.

It is difficult to quote reliable temperatures at which decalescence occurs. The temperatures vary with the amount of the carbon contained in the steel, and are much higher for high-speed than for ordinary crucible steel. Special electric furnaces are generally used for obtaining decalescence curves, but with care it can be done in an ordinary gas furnace, with a suitable pyrometer. All that is necessary is to bore a blind hole in a piece of the steel to be treated, to form a pocket to receive the end of the pyrometer. This must be of sufficient length to cover the resistance coil in the end of the pyrometer. The specimen should then be put into the furnace, with the pyrometer in, the gas applied, and, if the furnace is allowed to heat up very slowly toward a temperature of, say, 1380 degrees F. (750 degrees C.), the decalescence curve will be developed, if the pyrometer is a recording one. In the same way, if the furnace is allowed to cool slowly it will be seen that at the recalescence point, the specimen gives off heat and even increases in temperature for a time. Experiments of this kind are scarcely practicable for the average hardening shop, but when it is desired to find the lowest hardening temperature for a piece of steel, the magnet can be used to advantage.

Recapitulation. — To sum up, the decalescence point of any steel marks the correct hardening temperature of that particular steel. It occurs while the temperature of the steel is rising. The piece is ready to be removed from the source of heat directly after it has been heated uniformly to this temperature, for then the structural change necessary to produce hardness has been completed. Heating the piece slightly more may be desirable for either or both of the two following reasons. First, in case the piece has been heated too quickly, that is, not uniformly, this excess temperature will assure the structural change being complete throughout the piece. Second, any slight loss of heat which may take place in transferring the piece from the furnace to the quenching bath may thus be allowed for, leaving the piece at the proper temperature when quenched.

If a piece of steel which has been heated above its decalescence point be allowed to cool slowly, it will pass through a structural change, the reverse of that which takes place on a rising temperature. The point at which this takes place is the recalescence point and is lower than the rising critical temperature by some 85 to 215 degrees F. The location of these points is made evident by the fact that while passing through them the temperature of the steel remains stationary for an appreciable length of time. It is well to observe that the lower of these points does not manifest itself unless the higher one has been first fully passed. As these critical points are different for different steels, they cannot be definitely known for any particular steel without an actual determination. While heating a piece of steel to its correct hardening temperature produces a change in its structure which makes possible an increase in its hardness, this condition is only temporary unless the piece is quenched.

Quenching. — The quenching consists in plunging the heated steel into a bath, cooling it quickly. By this operation the structural change seems to be "trapped" and permanently set. Were it possible to make this cooling instantaneous and uniform throughout the piece, it would be perfectly and symmetrically hardened. This condition cannot, however, be realized, as the rate of cooling is affected both by the size and shape of the treated piece; the bulkier the piece, the larger the amount of heat that must be transferred to the surface and there dissipated through the cooling bath; the smaller the exposed surface in comparison with the bulk, the longer will be the time required for cooling. Remembering that the cooling should be as quickly accomplished as possible, the bath should be amply large to dissipate the heat rapidly and uniformly. Too small a quenching bath will cause much loss, due to the resulting irregular and slow cooling. To insure uniformly quenched products, the temperature of the bath should be kept constant, so that successive pieces immersed in it will be acted upon by the same quenching temperature. Running water is a satisfactory means of producing this condition.

The composition of the quenching bath may vary for different purposes, water, oil or brine being used. Greater hardness is obtained from quenching, at the same temperature, in salt brine and less in oil, than is obtained by quenching in water. This is due to a difference in the heat-dissipating power possessed by these substances. Quenching thin and complicated pieces in salt brine is unsafe as there is danger of the piece cracking, due to the extreme suddenness of cooling thus produced.

In actual shop work the steel to be hardened is generally of a variety of sizes, shapes and compositions. To obtain uniformity both of heating and of cooling, as well as the correct limiting temperature, the peculiarities of each piece must be given consideration in accordance with the points outlined above. In other words, to harden all pieces in a manner best adapted to but one piece would result in inferior quality and possible loss of all except this one. Each different piece must be treated individually in a way calculated to bring out the best results from it.

Theory of Critical Points. — The presence of the critical points in the heating and cooling of a piece of steel is a phenomenon. The most reasonable explanation is as follows:

While heating, the steel uniformly absorbs heat. Up to the decalescence point all of the energy of this heat is exerted in raising the temperature of the piece. At this point, the heat taken on by the steel is expended, not in raising the temperature of the piece, but in work which produces the internal changes here taking place between the carbon and the iron. Hence, when the heat added is used in this manner, the temperature of the piece, having nothing to increase it, remains stationary, or, owing to surface radiation, may even fall slightly. After the change is complete, the added heat is again expended in raising the temperature of the piece, which increases proportionally.

When the piece has been heated above the decalescence point and allowed to cool slowly, the process is reversed. Heat is then radiated from the piece. Until the recalescence point is reached, the temperature falls uniformly. Here the internal relation of the carbon and iron is transformed to its original condition, the energy previously absorbed being converted into heat. This heat, set free in the steel, supplies, for the moment, the equivalent of that being radiated from the surface, and the temperature of the piece ceases falling and remains stationary. Should the heat resulting from the internal changes be greater than that of surface radiation, the resulting temperature of the piece will not only cease falling but will obviously rise slightly at this point. In either event the condition exists only momentarily, but when the carbon and iron constituents have resumed their original relation, the internal heating ceases, and the temperature of the piece falls steadily, due to surface radiation.

Apparatus for Determining the Critical Points. — From the foregoing sections it is evident, first, that there is a definite temperature at which any carbon steel should be hardened, and, second, that a great loss occurs, both of labor and material, unless the hardening is carried out at this temperature. The actual shop problem thus presented is to determine readily and accurately the correct hardening temperature for any carbon steel that may be in use. This can be done by the use of various types of pyrometers; an apparatus made by the Hoskins Mfg. Co., of Detroit, Mich., is well adapted for the purpose. This apparatus consists of a small electric furnace in which to heat a specimen of the steel to be tested, and a special thermo-couple pyrometer for indicating the temperature of this specimen throughout its range of heating. The specimen itself should be properly shaped for clamping to the thermocouple.

The furnace may be operated on either alternating or direct current circuits. The furnace chamber is 21/16 inches in diameter and 2½ inches deep. Heat is produced by means of the resistance offered to the passage of an electric current through the "resistor" or heating element which in the form of wire is wound in close contact with the chamber lining. The furnace is designed so that it can be used on standard lighting circuits to which ready connection is made with a twin conductor cord and lamp plug. In operation, it consumes 3½ amperes at 110 volts, and is capable of producing a chamber temperature of 1830 degrees F., which is considerably higher than required for a carbon steel.

The pyrometer consists of a thermo-couple, connecting leads and indicating meter. The thermo-couple is of small wire so as to respond quickly to any slight variation in temperature. The welded end of this couple is slightly flattened to enable a good contact between it and the steel specimen. The meter is portable and indicates temperatures up to 2552 degrees F.

The specimen of the steel to be tested should be small, so as to heat quickly and uniformly. A well-formed specimen is made with two duplicate parts, each 1¼ inch long by ½ inch wide by ¼ inch thick. The pieces are clamped by means of two 1/8-inch bolts, one on each side of the welded part of the extreme end of the thermo-couple. Care is taken to form a tight contact, though not to cause an undue strain on the couple. The dimensions here given for the test specimen are not essential, though convenient; any pieces which will permit of tight contact with the thermo-couple and of heating in the furnace chamber may be used.

With the specimen fastened to the couple as just described the furnace is connected in circuit and the cover placed over the chamber opening. The temperature within the chamber rises steadily. When it becomes 1700 degrees F., the end of the couple, with specimen attached, is inserted in the chamber. The steel specimen rapidly heats, its temperature being constantly the same as that of the welded junction of the thermo-couple, due to the intimate contact between them. This temperature, indicated by the meter, will rise uniformly until the decalescence point of the steel tested is reached. At this temperature the indicating needle of the meter becomes stationary, the added heat being consumed by internal changes. These changes completed, the temperature again rises, the length of the elapsed period of time depending upon the speed of heating. With the furnace temperature kept nearly constant at the initial point, here given as 1700 degrees F., this "speed of heating" will be such as to allow of readily observing the pause in motion of the needle. The temperature at which this occurs should be carefully noted.

To obtain the lower critical point, the temperature of the piece is first raised above the decalescence point by about 105 degrees F. In this condition it is removed from the furnace and rested on top to cool. The decrease of temperature is at once noticeable by the fall of the meter needle. At a temperature somewhat below the decalescence point, varying with the composition of the steel, as previously mentioned, there is again a noticeable lag in the movement of the needle. The temperature at which the movement ceases entirely is the recalescence point. Immediately following there may occur a slight rising movement of the needle, as previously explained.

During these intervals of temperature lag, both during the heating and cooling of the steel, there may occur a small fluctuation in the temperature. In order to get results that are comparable, a definite point in each of these intervals should be considered each time a test is made. Hence, both the decalescence and recalescence temperatures are taken as the points at which the needle first becomes stationary. As all operations of heat-treatment of a steel center around its critical points, the importance of knowing these exactly is realized; to make certain, each test should be checked by a second reading. The time required for this is small. A close agreement of two succeeding readings will give assurance of the correctness of the determination.

Results Obtained from Sample Specimens. — In order to show graphically the necessity of quenching carbon steels at the proper temperature points, a series of specimen pieces of the same steel were treated at different temperatures. The steel used contained exactly 1 per cent carbon. A number of test specimens were made of this from adjacent parts of the same bar.

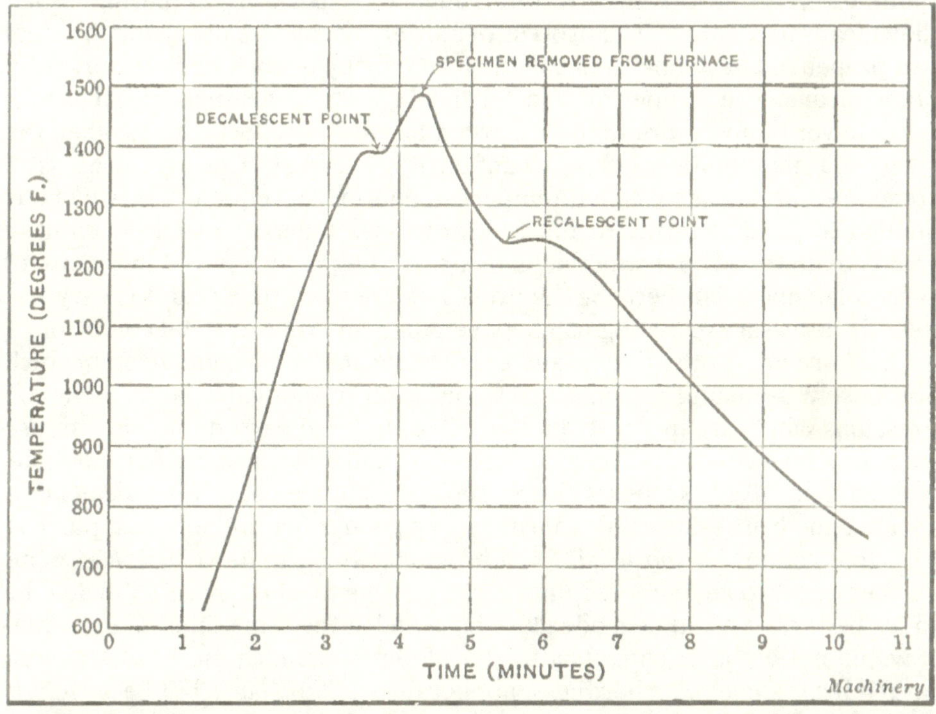

Fig. 2. Diagram showing the Relation between Time and Temperature when Heating Steel, & the Critical Temperatures of 1 percent Carbon Steel.

First the critical points of this steel were determined. Temperatures were recorded throughout both the heating and cooling. In the diagram, Fig. 2, these values have been plotted. The curve shows graphically the location of the critical points, and also the slight fall or rise of temperature as the case may be.

With this data obtained, seven specimens of the same steel were heated in the electric furnace, each to a different temperature. As these pieces were removed from the furnace they were immediately quenched in water. The temperature of the quenching bath was held constant at 45 degrees F. The hardened pieces were then broken at right angles and the fractured surface of each was photographed under a microscope. An inspection of the photographs at once showed the serious effects of overheating on the structure of the steel and hence on its strength.

One specimen was hardened just as the temperature reached the decalescence point. This showed clearly the direction in which the hardening takes place, namely, from the exterior toward the interior. This would naturally be expected as the temperature of the surface, which is exposed directly to the source of heat, reaches the critical point first. This condition indicates the necessity of heating the piece uniformly.

Tempering. — Although the mistake is rarely made in the hardening-room itself, it is surprising how often engineers speak of "tempering" when they mean "hardening." The process of hardening consists of raising the steel to a proper red heat, and then cooling it rapidly by quenching in water, or by other means. The tempering is a further process of heating the steel to a much lower temperature than for hardening, and the effect is to toughen the steel, and also, unfortunately, to soften it somewhat. The first process is sometimes erroneously called tempering, and more frequently this word is applied in general terms to the two processes together, but it is entirely wrong to do this. The distinction is made not simply by a purist in the use of correct language, but because the processes are so distinct that it is very frequently necessary to distinguish between them in order to avoid confusion.

There are many other terms and expressions in the hardening-room which are loosely used, and in some cases confusion of thought results. Some expressions which are incorrect in themselves have so passed into current use that they must, perhaps, be tolerated and made the best of. For example, when a man says that his tools are "too hard" he almost always means that they are too brittle. Surely the hardness is a good thing, and for most purposes a tool cannot be *too hard*. Even a punch, which under existing circumstances ought to be tempered until it can just "be touched by a file," would be much better if it could be made glass-hard and at the same time so tough that it would not break or chip. As a matter of fact, the punch must be tempered until it loses some of its hardness, not because it is *too hard,* but because it is *too brittle*. It would be better if the brittleness could be removed without impairing the hardness. The loose expression has established itself because the ideas of hardness and brittleness are so intimately associated in men's minds that frequently they can hardly be dissociated. There are times, however, when the expression which is so frequently misused is really required in its correct sense; for example one sometimes uses a lead or copper hammer instead of a steel one, because the steel hammer would be *too hard* and would bruise the work.

The Use of Pyrometers. — The hardening of carbon steels for highest quality and greatest saving entails, then, three things. First, a definite knowledge of what constitutes the correct temperature at which to harden the steel. The second point necessitates a positive means of accurately determining this hardening temperature for any carbon steel. The third consideration is that the correct hardening temperature, once determined, is actually carried out in the hardening work. A simple and effective way of doing this is by checking the temperature of the hardening furnace by means of a pyrometer. When there is a large quantity of work to be hardened, economy dictates a permanent installation of pyrometers. The convenience of such installations is manifest. A thermo-couple is placed in each furnace. A number of these, from three to sixteen, depending upon individual conditions, are connected by wire leads, through a selective switch to one meter. By a turn of the switch, the temperature of any furnace may be read at once from the meter. This makes it possible for the foreman to know definitely at a single point, the temperatures of all of the hardening furnaces in use.

The Brayshaw Experiments on the Heat-treatment of Steel. — Having now reviewed the principles underlying the hardening of steel, we are ready to take up in greater detail the processes and devices used for obtaining the required results. Before doing so, however, a brief account will be given of a remarkable series of experiments that have been made to ascertain, as definitely as possible, the laws that govern the heat-treatment of steel.

The hardening of steel has always been considered an operation for which definite rules could not be laid down, but in which the experience and judgment of the hardener would almost exclusively have to be relied upon. Practically the only definite rules that have been laid down are that steel should be hardened at as low a heat as possible, and that there is a definite temperature for each kind of steel above which it must be heated to harden at all.

Apart from this meager information, a few generalities only have been furnished for the guidance of men doing this work, including directions for cooling, so as to minimize the risk of cracking; but definite information on the process of hardening is singularly lacking, and many have considered it impossible to shape rules for this operation, even as the result of careful experiments.

For this reason the experiments conducted by Mr. Shipley N. Brayshaw of Manchester, England, the results of which were reported in a paper read before the Institution of Mechanical Engineers at the April 15, 1910, meeting, and which are recorded in detail in the following, are all the more remarkable and of great interest to everyone engaged in mechanical work. These experiments appear to have been made with extraordinary care, and two of the points brought out in his investigations deserve to be particularly mentioned. One of them relates to the shortening or lengthening of steel in hardening. It has often been stated that steel is unreliable and not uniform in this respect, and that the same kind of steel has sometimes been known to shorten in hardening, and sometimes to lengthen. This phenomenon has been attributed to defects, or, at least, to special conditions in the steel, over which the hardener has no control. Mr. Brayshaw's experiments, however, show that the shortening and lengthening of steel in hardening follows a uniform law, and that steel hardened below a given temperature, which he calls the "change-point," will shorten when quenched; whereas the same steel, if heated above this change-point before quenching, will lengthen. He also shows that by heating the steel in two furnaces, first bringing it up to a certain temperature in one, and then soaking it for a given time at a definite temperature in the other, it is possible to harden steel so that it will neither lengthen nor shorten when quenched. This is, without doubt, the first definite information that has been recorded on the change in the length of steel in hardening, the uncertainty of which has caused considerable difficulty in making accurate taps, dies, etc.

Another interesting point brought forth by Mr. Brayshaw's experiments relates to the proper hardening temperature. While it is possible to harden steel within a temperature range of about 200 degrees F., and obtain what to the ordinary observer would seem to be good results, the *best* results are always obtained within a very narrow range of temperatures, approaching closely the decalescence point, or the temperature at which steel changes into a condition when it can be hardened by quenching. It is interesting to note that this result agrees with the old theory that steel should be hardened at as low a temperature as possible. In a number of cases certain tools are found to last exceptionally well, while other tools in the same lot show only ordinary durability, although no difference can be detected in the grain of the hardened steel. Such differences can now be accounted for by the slight variations in the temperature at which the various tools have been hardened.

These experiments open up an entirely new field for investigation and new possibilities in the hardening of tool steel. They indicate that in cases where it is important to prevent the cracking of tools in hardening and the deformation of tools due to internal stresses, they should be hardened at a temperature higher than that which gives the best results as regards hardness only. Thus we find that certain of the desirable qualities in a hardened tool are antagonistic; that is, we are unlikely to obtain a tool having extreme hardness and elastic limit and which at the same time is not likely to crack or lose its shape. Under these conditions one of the necessary qualities must be partly sacrificed to another, and a tool must be hardened so as to obtain good general results rather than the best specific one.

The influence of previous annealing on hardening is also of interest; and many of the points brought out by the results of these experiments are well worth considering. The experiments deal exclusively with the results obtained from two kinds of carbon tool steel that, except for minute variations, differed only in the fact that one of them contained about 0.5 per cent of tungsten. The steel contained on an average of 1.16 per cent carbon, 0.15 per cent silicon, 0.36 per cent manganese, 0.018 per cent sulphur, and 0.013 per cent phosphorus. The whole work of investigation was devoted to questions directly connected with machine shop hardening, with the aim in view of throwing light on the many problems met with in daily practice.

Hardening Temperatures. — The hardening point of both low-tungsten and carbon steel may be located with great accuracy, and the complete change from soft to hard is accomplished within a range of about 10 degrees F., or less. After the temperature has been raised more than from 35 to 55 degrees F. above the hardening point, the hardness of the steel is lessened by further increases in the temperature, provided the heating is sufficiently prolonged for the steel to acquire thoroughly the condition pertaining to the temperature. There is a "change-point" at about 1615 degrees F. in low-tungsten steel and at a somewhat higher temperature in carbon steel. One of the several indications of this change-point is the shortening of bars hardened in water at temperatures below that point, whereas the bar lengthens if this temperature is exceeded at the time of quenching. Practically the same results are obtained by heating low-tungsten bars to any temperature from 1400 to 1725 degrees F. and quenching in oil, as by quenching in water.

Length of Time of Heating. — Regarding the effect of heating to various temperatures for various lengths of time before quenching for hardening, the following conclusions are drawn: Prolonged soaking up to 120 minutes at temperatures at which the hardening change is half accomplished in 30 minutes, does not suffice to complete the change. Prolonged soaking for hardening at a temperature of 1400 degrees F. has a slightly injurious effect on the steel, but does not materially influence the hardness. At a temperature of about 1490 degrees F. a great degree of hardness is attained by quick heating, but the hardness is impaired with 30 minutes soaking. Prolonged soaking for hardening at a temperature of about 1615 degrees F. has a seriously injurious effect upon the steel. A specially great degree of hardness may be obtained by means of soaking at a high temperature, as at 1615 degrees F. for a very short time, but even as long a time as 7½ minutes is long enough to seriously impair the hardness.

The temperature of brine for quenching is of considerable importance. Both low-tungsten and carbon steel bars quenched at 41 degrees F. were decidedly harder than bars quenched at 75 degrees, and quenching at 124 degrees F. rendered the bars much softer.

Effects of Previous Annealing. — The method of previous annealing affects the hardness of steel considerably. The elastic limit of low-tungsten bars hardened at either 1400 or 1580 degrees F. varies according to the annealing they have undergone. The elastic limit is high after annealing at about 1470 degrees F. for 30 minutes, or 1290 degrees F. for 120 minutes, but it is seriously impaired by annealing at 1470 degrees F. for 120 minutes.

If low-tungsten steel is annealed at 1725 degrees F. and hardened at 1400 degrees F., the elastic limit is inferior, and the adverse effect of the previous annealing is much more pronounced if the hardening is done at 1580 degrees F. The elastic limit of carbon steel annealed at any temperature between 1290 and 1725 degrees F. and hardened at either 1400 or 1580 degrees F. does not vary by nearly such great amounts as the elastic limit of the low-tungsten bars, and the highest annealing temperature given above is not injurious so far as the elastic limit is concerned.

The hardness of low-tungsten bars hardened at 1400 degrees F. decreases from a high scleroscope figure to a low one as the temperature of annealing increases from 1290 to 1725 degrees F. The hardness is increased by prolonging the annealing at the lower temperature. The hardness of low-tungsten steel hardened at 1580 degrees F. is fairly constant at a moderately high sceleroscope figure, whatever the temperature of annealing.

Effect of Heating in Two Furnaces. — An interesting part of the experiments relates to the use of two furnace heats for hardening, heating the steel first in one furnace to a certain temperature for a given time, and then immediately, without cooling, soaking in a second furnace at a known temperature and for a definite time. These experiments show that low-tungsten and carbon steel bars heated for half an hour to temperatures between 1545 and 1650 degrees F. are not much affected so far as their elastic limit and maximum strength are concerned by a further immediate soaking for half an hour at 1400 degrees F. If, however, the temperature in the first furnace is 1725 degrees F., the low-tungsten steel is much improved by a further soaking at 1400 degrees F., but the carbon steel is much injured by the same treatment. Bars of low-tungsten steel heated for 30 minutes, at 1615 degrees F. and then soaked at 1332 degrees F. for a further 30 minutes, give a high elastic limit and maximum strength, and are harder than if the second soaking were at a temperature of 1400 degrees F. The carbon steel, again, is but little affected by these variations in the second furnace.

The change of length in hardening, however, of both low-tungsten and carbon steel is much affected by the above variations in the temperature of the second furnace. Good results as regards elastic limit and maximum strength, and also as regards hardness, are obtained by very short soaking, first at a high temperature, say 1615 degrees F., and then at a low one, the results being best when the second temperature is near to or a little below the hardening point. If the furnace be at a sufficiently high temperature it is easy either by variations of the temperatures of the two furnaces, or by variations in the time of soaking, to arrive at a treatment of the steel, both low-tungsten and carbon, whereby they neither lengthen nor shorten. Under the same treatment carbon steel has -a greater tendency to shorten than low-tungsten steel.

Miscellaneous Results. — Other experiments showed that low-tungsten steel heated to 1580 degrees F. for 15 minutes and quenched in oil has a higher elastic limit and is harder than carbon steel similarly treated. As to annealing, it was found that bars annealed at a temperature of 1470 degrees F. or below became slightly shorter by the annealing process, and this action was more pronounced in the case of carbon steel than tungsten steel. Annealing at a temperature of 1650 degrees F. causes both low-tungsten and carbon steel to lengthen.

It was found that recalescence of low-tungsten steel takes place gradually at a temperature of 1348 degrees F., and more readily at 1337 degrees F., and further that the recalescence at either of the above temperatures is very much retarded if the steel is cooled from a maximum heat of 1634 degrees F.

Regarding hardening cracks, it is shown that both for low-tungsten and carbon steel, such treatment as produced the highest elastic limit accompanied by the greatest hardness is frequently the most risky. The risk of hardening cracks is reduced if the steel is heated for a sufficient length of time to a temperature of 1650 degrees F. or a little above. Low-tungsten steel is more liable to crack in hardening than is carbon steel.

Effect of Tempering. — Tempering experiments showed that little effect was produced by the tempering of carbon steel to 300 degrees F. for 30 minutes. Tempering the same steel to 480 degrees F. for 15 minutes, however, caused it to soften considerably and to shorten in length. For low-tungsten steel the elastic limit was increased considerably by tempering up to a temperature of 480 degrees F. The maximum strength of the same steel coincides with the elastic limit for bars either untempered or tempered at 300 degrees F. for 15 minutes, but it then rises rapidly with further tempering. The hardness, as measured by the scleroscope, was considerably reduced by tempering at 300 degrees F. and still more at 390 degrees F., but was not so much affected by further tempering at 480 degrees F. The length of the low-tungsten bars was reduced by tempering up to a temperature of 480 degrees F.; the higher the temperature, the greater was the reduction in length.

Effect on Tensile Strength. — The following conclusions refer to low-tungsten steel, but there is no reason to doubt that they are also applicable to carbon steel. A variation in the hardening temperature of only 9 degrees F., the extremes being respectively above and below the proper hardening temperature or decalescence point, had a tremendous influence on the extension under load, but the maximum strength of the bars so treated did not differ much. A very good bar was produced by quenching from a temperature fully 108 degrees F. above the hardening temperature. A heat of only 5 minutes' duration produced a harder bar than a heat of 25 minutes, the maximum temperature in both cases being 1470 degrees F., or a little above; but the bar heated for a shorter time gave a much lower elastic limit. The maximum strength alone is not necessarily any indication of the condition of the steel in question, or of the treatment to which it has been subjected; nor is the hardness alone necessarily an indication of the condition of the steel or the treatment.

The following conclusions refer both to tungsten and carbon steels. Tempering up to a temperature of 570 degrees F. gradually increases the maximum strength and the elastic limit, although some irregularities enter which have not been fully accounted for. Tempering to this temperature reduces, for a given stress, the extension under load and the permanent extension.

In conclusion it may be stated that these experiments show that steel of the quality treated in these experiments may be hardened within a temperature range of about 215 degrees F. The lower end of this range is very sharply defined, but the highest temperature allowable is difficult to determine, and as far as the appearance of the fracture is concerned there is but little evidence of improper hardening until the temperature of the proper hardening point has been exceeded by 270 degrees F. So wide, in fact, is the margin of allowable variation for hardening that when the hardness is decided by the appearance of the fracture alone, any workman of average skill can easily keep within the limits and judge the temperature by sight alone, and as a matter of fact this is being done all the time in the manufacture of such articles as pocket knives, small files, etc., which are hardened by the thousands with practically no waste. But, of course, it must not be understood that articles so hardened reach anything like their maximum efficiency, because even small variations in the heat-treatment previous to the quenching have a pronounced effect upon the condition of the steel, and even the previous treatment, such as the annealing to which the steel has been subjected, may influence the final result.

While it is thus easy to harden so as to obtain reasonably *good* results, the production of the *best* results necessitates a high degree of accuracy which can never be obtained by sight alone, and it is also important to notice that the difference between good hardening and the best hardening is very great. As an example may be mentioned the hardening of razors. It is sometimes said that whatever price one pays for a razor, the buying is a game of chance. Occasionally one hears of a remarkable razor that holds its edge as if by magic, while others of the same make and type may not be anywhere near as good. All of them, however, would show to the eye practically the same fracture, and apparently seem to have been treated in the same way. The experiments referred to above, however, indicate that there may have been a slight difference in the hardening temperature and consequently in the subsequent condition of the steel, and also that it would be possible to harden every razor in a gross so that each one would be truly a duplicate of the best. The same, of course, holds true of a great number of other tools.

There have, for some years, been efforts made by steel makers to discover new alloy steels, and splendid success has been obtained in this direction, but there is still a wide field for the steel users for discovering the best use of the material already known. It is of little avail that occasionally tools show marvelous results, unless the hardener can at any time produce the same results with the same steel. The time is likely to come when all the factors in the hardening of tool steel will be controlled with accuracy within predetermined limits, and any failure may be investigated and the cause located with as much certainty as if a mistake had been made in the machine shop.

Carbon Steel vs. High-speed Steel. — The idea that the proper hardening of carbon steel will make it possible to obtain better results by the use of this steel than has ordinarily been the case in the past has been expressed in this country as well as in England. At least one large machine tool building and tool-making firm has made extensive inquiries in the direction of determining proper methods for hardening ordinary carbon steel so as to obtain the best results, and several writers on mechanical subjects who have investigated the subject concur in the opinion that carbon steel, properly heat-treated, can be made to produce better results than is usually expected. One writer in Machinery gave voice to this opinion as follows:

"High-speed steel is in fashion nowadays. This fact together with the high degree of skill required to get the best results from carbon steel has caused this steel to be neglected. For some kinds of work, however, carbon steel is superior to any high-speed steel on the market, if it is dressed properly and receives the proper heat-treatment.

"The carbon in steel may be in one of two forms: Annealed steel has the carbon in the non-hardening or cementite form. Hardened steel has the carbon in the hardening or martensite form. That these are two distinct forms may be seen by taking a small piece of annealed steel and a small piece of hardened steel and dissolving each in hydrochloric acid. The annealed steel will dissolve, leaving a black residue, while the hardened piece dissolves, leaving no residue. This shows that in one case some of the carbon is in the free or graphitic form and does not dissolve in acid, while in the other case, the carbon is in the combined form and all dissolves."

Regarding the use of the magnetic needle for indicating the correct hardening temperature, the same writer says:

"There are two hardening methods which depend on the fact that steel loses its magnetic properties when the hardening point is reached. A piece of steel heated to a dull red and brought into the plane of a magnetic needle will attract the needle. If heated until the temperature is above the hardening point, there will be no attraction and the needle will not be affected by the presence of the steel. In using this test, care must be taken that the presence of the tongs does not mislead the workman. The comparatively cool tongs may attract the needle even when the steel is above the critical point. An ordinary horse-shoe magnet may be used instead of the magnetic needle. It is less likely to mislead because of the tongs or cooler parts of the steel, but is less sensitive. A bar magnet hung on a pivot at the center and provided with a handle can be used very satisfactorily. It can be introduced into the furnace to test the steel during the process of heating and is more convenient than either of the other methods. It is not necessary to test every piece of steel, but a test should be made whenever the person takes another grade of steel or whenever the light changes. The intensity of the light makes a great difference in the color of a piece of iron or steel at a given temperature."

General Rules for Hardening. — If steel workers would observe the following in all cases when hardening steel, they would have better results: *Harden carbon steel at the lowest possible heat and always on the rising heat.* The last part of this rule is the one more often overlooked. Steel may be forged at a higher heat than the hardening heat, but should in all cases be annealed before being heated for hardening. The grain of the steel corresponds to the highest heat it has received since it was black. If a piece of steel is forged at 1600 degrees F. and allowed to cool down to 1400 degrees F. to harden, it will have a grain corresponding to 1600 degrees F.

Chapter Two - Heating the Steel for Hardening

The furnaces used in hardening and tempering are heated either by gas, oil, electricity or solid fuels, such as coal and coke. Furnaces using oil or gas are made in many different styles and sizes to suit various classes of work, but differ very little in their general arrangement.

Simple Type Gas Furnace. — Fig. 1 shows at *B* and *C* a very simple type of crucible furnace in which gas is burned under moderate air blast. There should be not less than two burners when this arrangement is used. The section *B* shows the arrangement of these burners in the vertical plane and the section *C*, their arrangement in the horizontal plane. With this arrangement of burners, the flame will travel several times around the crucible instead of passing as nearly as possible in a straight line from the burner to the vent, as would be the case if the furnace were constructed as at *A*, which is an objectionable design.

At *D*, in the same illustration, is shown a section of a burner which can be made up from standard pipe fittings. If the tip burns out, as it probably will in time, the flanged joint enables it to be easily removed, and the burned parts can be cheaply and quickly replaced. Between the flanges, a gasket of asbestos paper or similar material should be used. For a furnace of the dimensions indicated at *A*, fittings for 1½-inch wrought-iron pipe will be large enough, if two burners are used, and a 2-inch pipe will be sufficient to bring the gas and air from the mixer to the tee supplying the burners. If the air used carries a pressure of forty pounds per square inch or more, the air should, for the sake of economy, be used in a jet blower. A design of a combined jet blower and mixer which can easily be designed from pipe and fittings is shown in Fig. 2. The fittings should not be smaller than 2-inch pipe size for the cross, 1½-inch pipe for the gas, and 1/8- or ¼-inch pipe for the compressed air. The pressure beyond the mixer need not be more than two pounds per square inch, and often one pound is sufficient.

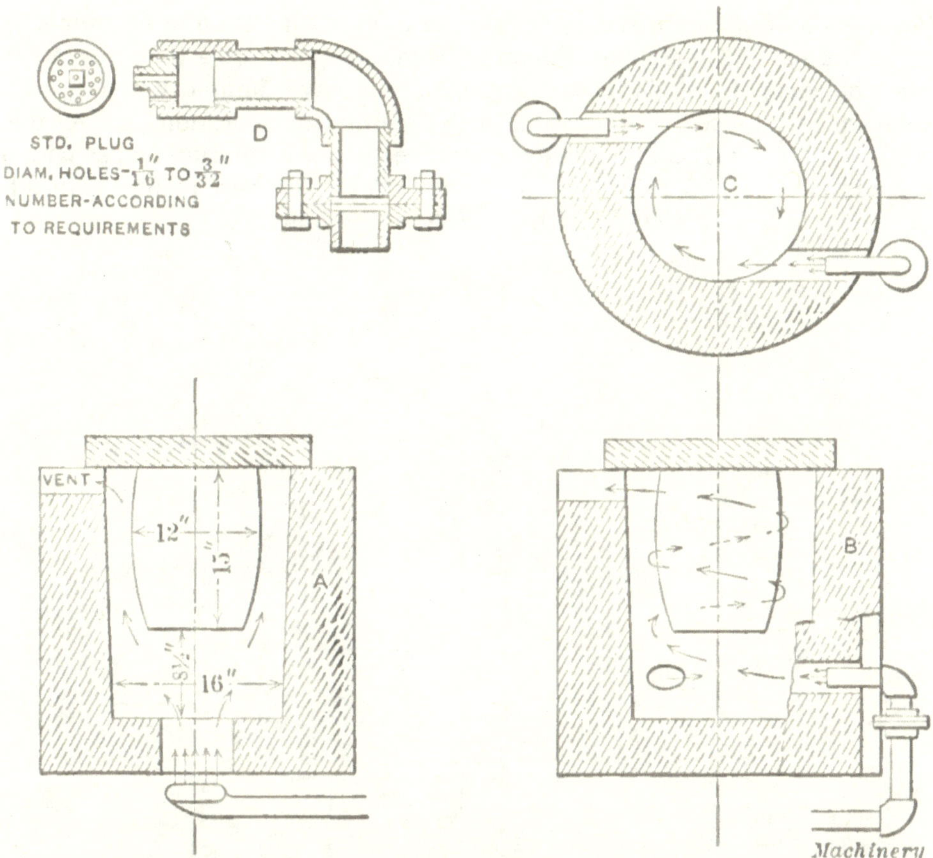

Fig. 1. Simplest Type of Crucible Gas Furnace showing Objectionable and Approved Designs

Gas furnaces use either natural, artificial or producer gas. Some gas furnaces are equipped with an automatic apparatus which operates in conjunction with a pyrometer for controlling the temperature to within a few degrees of a given point. The air supply is generally obtained from a positive blower, although when a compressor is installed for operating pneumatic tools, the air is sometimes utilized for the furnaces by interposing reducing valves to diminish the pressure. Artificial gas is more expensive than oil, but is cleaner, and the installation of supply tanks, such as are required for oil, is avoided. Producer gas obtained from a separate plant is not economical unless there is a considerable number of furnaces. When oxidation or the formation of scale is particularly objectionable, furnaces of the muffle type are often used, having a refractory retort in which the steel is placed so as to exclude the products of combustion. These muffles must be replaced very frequently and more fuel is required than when an oven type of furnace is used.

Oil-burning Furnaces. — The use of oil in furnaces for the heat-treatment of steel possesses, in many cases, certain advantages over other methods of heating. Chief among these advantages is the consideration of economy, as oil in the past, at least, has generally been cheaper to use than any other available form of fuel. The consideration of economy is limited, however, by a somewhat increased complication in the method of operation, and on this account, oil is not recommended for furnaces that will operate on gas with a consumption of 230 cubic feet per hour or less. The best results with oil-heated furnaces are secured with the larger-sized units.

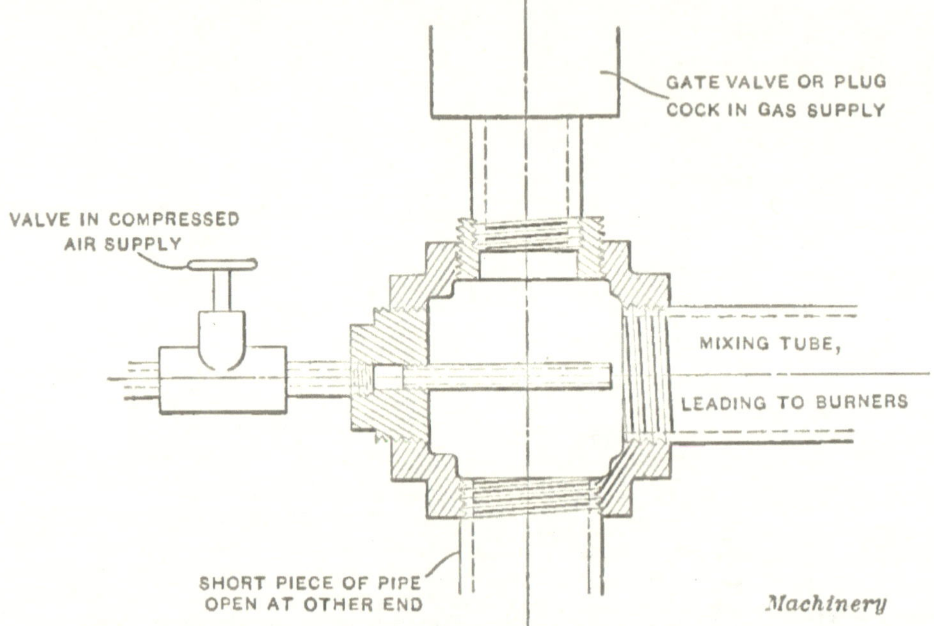

Fig. 2. Combined Jet Blower and Mixer made from Pipe Fittings

Before entering upon a description of any specific type of furnace, it will doubtless be advisable to give a brief description of the method on which oil furnaces operate. The oil is fed to the furnace from a tank underground, a pumping system being used to maintain a pressure of about 40 pounds per square inch in the supply pipe. The oil emerges from the burner through an orifice about 1/16 inch in diameter. This orifice is surrounded by a second pipe through which steam or air is supplied under pressure. The fine stream of oil is taken up by this stream of compressed air or steam and "atomized" or broken up into very finely divided particles. The oil mixed with the air in this way forms the combustible mixture that is burned in the furnace.

In those cases where the oil is atomized by steam, it is necessary to supply a certain amount of additional air to get the desired combustion. To understand the way in which this combustion proceeds, it must be understood that steam is a chemical compound consisting of two parts of hydrogen and one part of oxygen. When the steam impinges upon the white hot brickwork of the furnace, the chemical union is broken, hydrogen and oxygen being liberated. The oxygen set free in this way is used in effecting the combustion of the oil, and the hydrogen is carried into the furnace. Now hydrogen is itself a combustible gas, and is burned in the furnace by the oxygen of the additional air which is supplied for this purpose. This combustion of hydrogen takes place further from the burner than the point at which the bulk of the oil is burned, and helps considerably in maintaining a uniform temperature. When the oil is atomized by a stream of compressed air, there is no hydrogen present to be burned in the furnace.

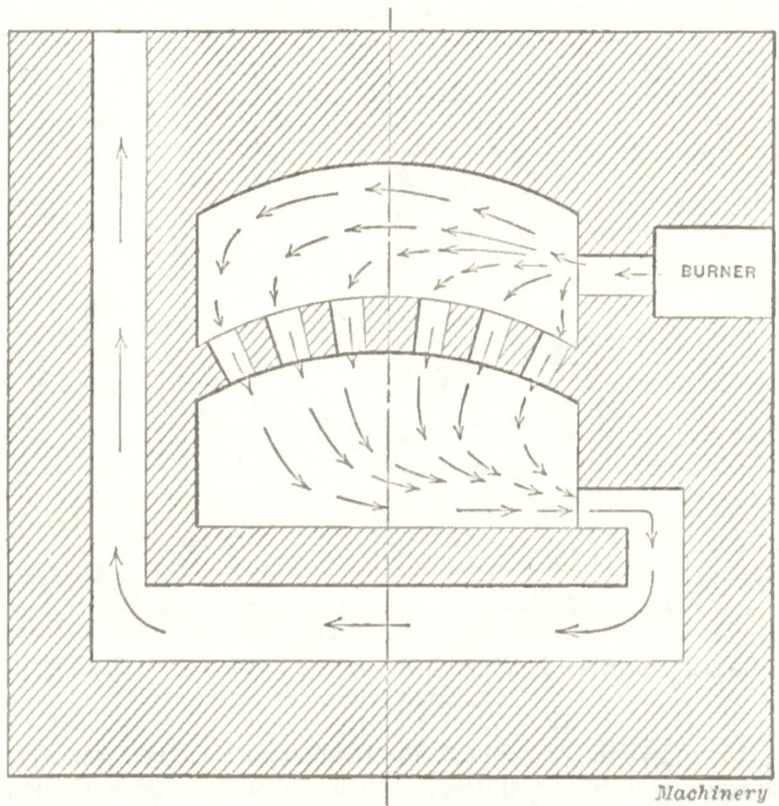

Fig. 3. Diagrammatical Section showing the Principle of the Over-fired Type of Furnace

Types of Oil-burning Furnaces. — Crude oil and kerosene are commonly used in oil-heated furnaces. To insure an unvarying temperature, the air and fuel pressures should be uniform. Two general types of oil-fired furnaces are shown in Figs. 3 and 4. That shown in Fig. 3 is of what is called the over-fired type, in which the atomized gas from the oil burner passes into a space above the heating chamber, from which it is separated by an arch perforated by a number of openings through which the burning gas passes. This arrangement gives a uniform heat in the working chamber and the temperature is easily controlled. With a proper fuel supply and system of control, the temperature can be maintained within 25 degrees F. at any predetermined point between 600 and 1800 degrees F. Tests lasting for 1½ hour have been made in which the variation of temperature did not exceed 10 degrees.

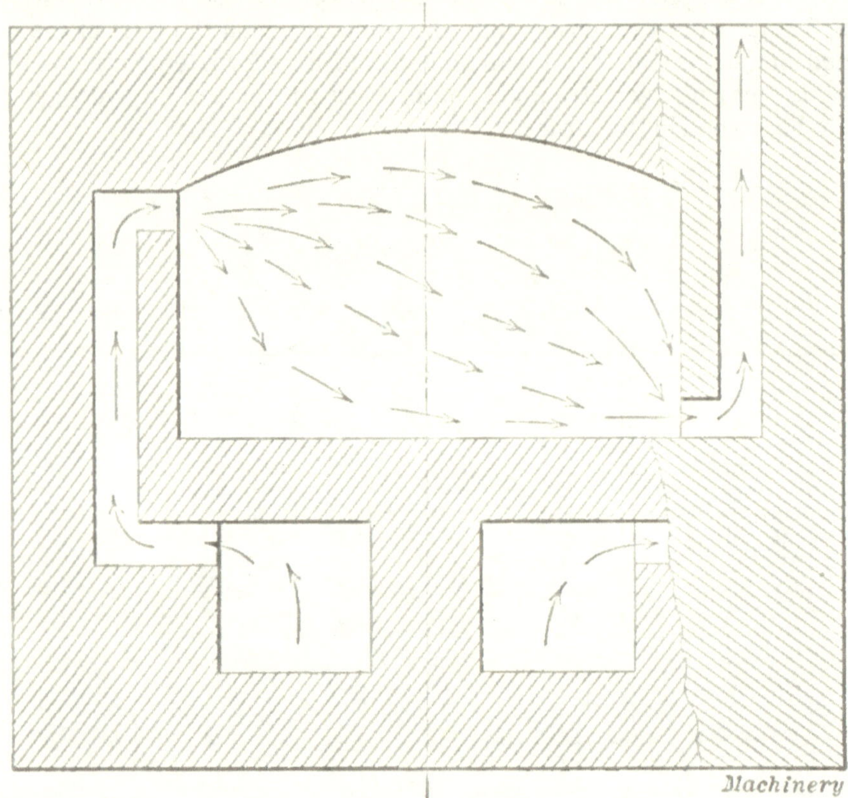

Fig. 4. Diagrammatical Section showing the Principle of the Under-fired Type of Furnace

In Fig. 4 is shown the principles of a so-called under-fired furnace, in which the atomized gas from the oil first passes into chambers beneath the heating chamber. The combustion takes place in these lower chambers and the gas then passes through flues into the top of the heating chamber, from which it passes to the outlet flue, as indicated. It will be seen that the construction of a furnace of this type is simpler than is that of the over-fired type, so that the first cost is less. The cost of repairs is also smaller, but the fuel consumption is slightly greater and it is more difficult to maintain a uniform heat in all parts of the heating chamber than in the case of an over-fired furnace. When built in smaller sizes, however, the heat is easily controlled, and a furnace of this construction is suitable for tool hardening and tempering.

Characteristics of Fuel Oils. — In an effort to determine the calorific values in British thermal units per pound of fuel oils, sixty-four samples of petroleum oils ranging from heavy crude oil to gasoline, representing the products of the principal oil fields in the United States, were examined for calorific power by combustion in oxygen in the Atwater-Mahler bomb calorimeter. It was found that the oils varied in fuel value from about 18,500 to 21,100 B.T.U. per pound. In general, the decrease in calorific power with an increase in specific gravity is quite regular, so that the relation between the specific gravity and the heat value may be expressed approximately by means of a simple formula, as follows:

B.T.U. = 18,650 + 40 (Number of Degrees Baumé — 10).

When the heat values calculated from the specific gravity by means of this formula were compared with those actually determined by experiments, it was found that in one-ninth of the cases the difference was greater, and in eight-ninths it was less than one per cent. Hence, the heat value of commercially pure petroleum oils can be determined by the density of degrees Baumé with sufficient accuracy for most practical purposes.

The heat value of oil is reduced by the presence of small percentages of water. Therefore, if the oil contains water, it should be passed through a filtering tank before going to the burners. In this filtering tank the water settles to the bottom and can be easily drawn off. The oil should be heated before going to the filtering tank, as the water in the oil is more easily separated out of hot oil than cold oil, first, because heated oil offers less resistance to freeing the water, and, second, because there is a greater expansion of oil than water due to the heat, and the water, therefore, has a relatively greater specific gravity.

Kerosene for Steel Heating Furnaces. — During the last few years the price of fuel oil has steadily advanced and in the fall, 1913, the oil refineries, in a number of states, notified industrial plants that they would be unable to supply fuel oil after a certain date. At the present time it is almost impossible to enter into a contract with the refineries for a year's supply of this oil. This condition has caused many manufacturing plants to analyze the situation and try to find a substitute for fuel oil without installing new equipment in the heat-treating department. The Continental Motor Mfg. Co., Detroit, Mich., has made numerous experiments with different kinds of oils and burners in order to be ready to meet the situation in case there will be a permanent shortage of fuel oil. Kerosene has proved most satisfactory, after many experiments, and the details given in the following show the results that have been obtained.

The heat-treating room is of the most modern design and contains many unique features which are of interest. This department is housed in a fireproof building of structural steel and metal sash, and is entirely isolated from the other buildings. All auxiliary apparatus, such as oil and circulating pumps, air compressor and storage and cooling tanks, is located in the basement, so that the entire first floor is given up to the furnaces and quenching tanks. The furnaces were made by the American Shop Equipment Co., and consist of four double-chamber case-hardening furnaces of semi-muffle type, with heating spaces 54 by 27 inches and 18 inches high. The combustion chambers are 27 inches wide by 7½ inches high. The two burners in front of each chamber are supplied with oil at 18 pounds pressure and air at 1½ pound pressure.

The furnaces show a very even temperature, and a laboratory pyrometer fails to show over 10 degrees variation in any part of the heating chambers. Each furnace is equipped with both an indicating and a recording pyrometer, so that the operator can tell at a glance the temperature of each chamber. To facilitate the handling of material, the hearths of all furnaces, packing tables, trucks, quenching tanks and other equipment are made the same height from the floor. The door openings are made the full width of the heating chamber and are counterweighted.

The quenching tanks are of special design and are located in the center and opposite the furnaces so that they are readily accessible. Water, brine and oil are used for quenching, and by means of cooling tanks and circulating pumps the quenching mediums are kept at a constant temperature. The pumps deliver the liquid in the bottom of the tanks by means of perforated pipes which are protected by a wooden grating. This wooden grating also acts as a cushion for any parts which happen to fall into the tanks. The entire surface of the liquid is removed at a uniform rate by means of numerous outlets located near the top of the tank and connected with a common overflow pipe. The rate of flow can be governed by means of a valve fitted to the inlet pipe near the floor. The oil quenching tanks have a cover which can be closed very quickly in case the oil ignites. The quenching mediums are circulated through coils and are cooled by means of cooling water which is varied to suit conditions. The circulating pumps are direct motor-driven centrifugal pumps, located in the basement. The liquid is fed by gravity to these pumps, which does away with the troublesome feature of priming.

The oil is fed to the burners by direct-connected, motor-driven rotary pumps, in duplicate. The oil is pumped from a 12,000-gallon tank located outside the building, and the surplus oil is returned to the storage tank by means of a relief valve and overflow. This system allows a constant pressure of oil at the burners at all times, and the fuel oil is entirely drained back to the storage tank when the heat-treating room is not in operation. The elevation and plan, Figs. 5 and 6, show the storage tank and arrangement of the furnaces, oil and brine cooling coils, quenching tanks, etc.

A General Electric turbo air compressor furnishes the necessary air for the burners at 1½ pound pressure. The air lines were carefully laid out in order to reduce the friction to a minimum. The advantage of this type of air compressor is the constant pressure of air, regardless of the volume.

Tests were repeated for ten successive days on both kerosene and fuel oil. Both oils burned uniformly and needed very little attention after the proper regulation. One of the chambers of a double furnace was used for these tests, and in order to accurately measure the oil, a tank of 60 gallons capacity was

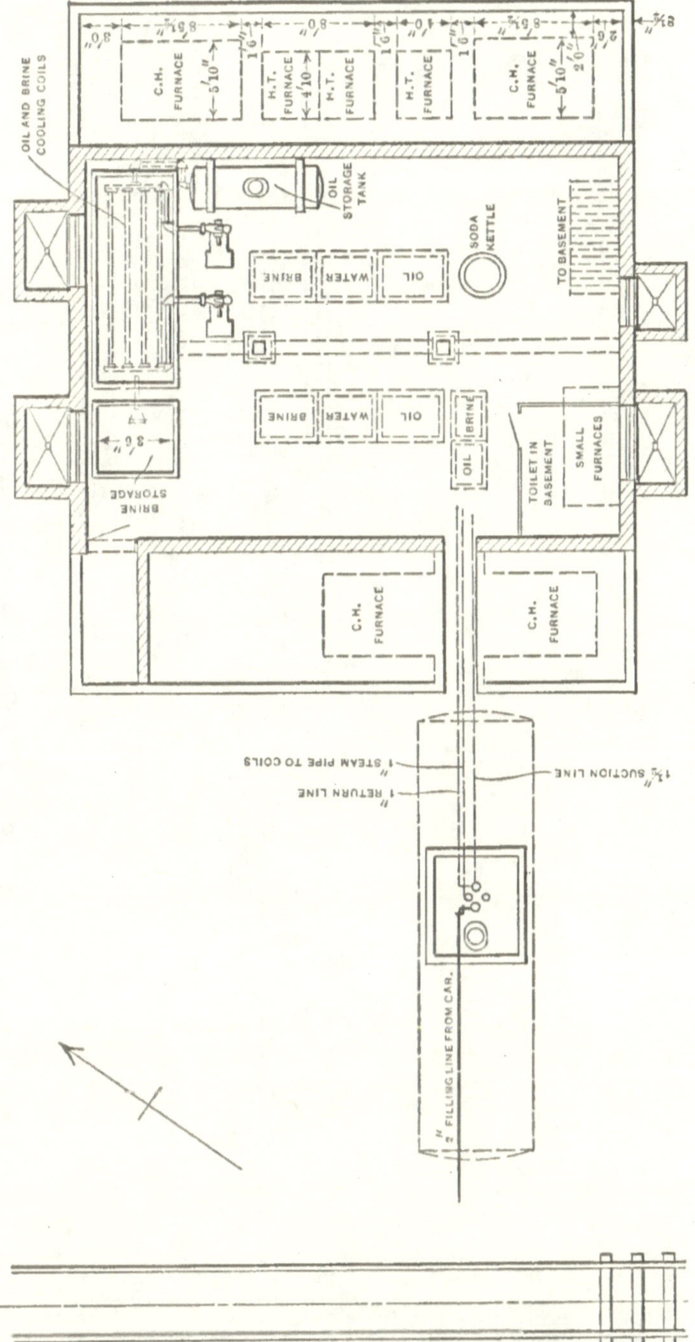

Fig. 5. Plan showing the entire installation at the Continental Motor Mfg. Co., Detroit, Mich., for burning kerosene in oil-burning furnaces. Illustration shows the storage tank and arrangement of the furnaces, oil and brine cooling tanks, quenching tanks, etc.

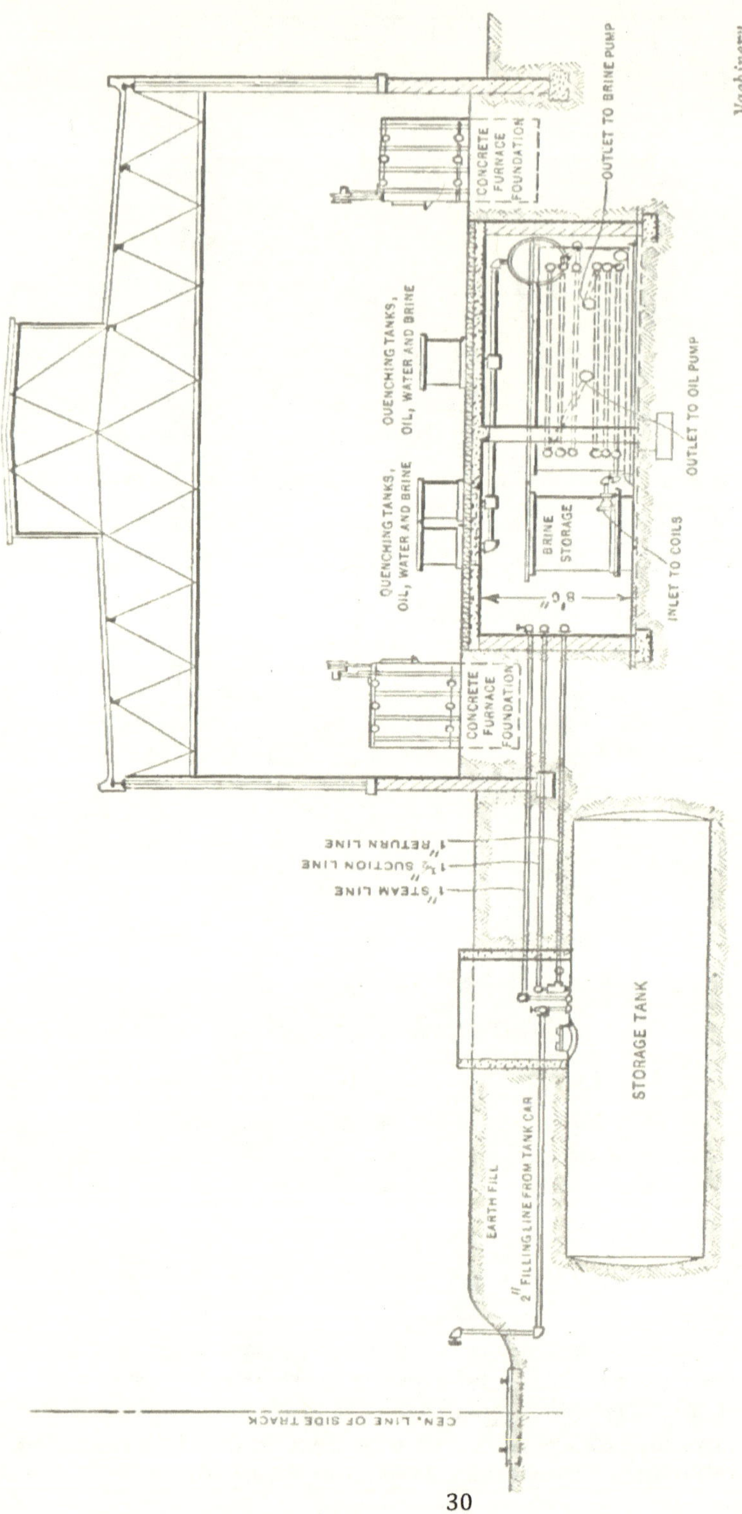

Fig. 6. Elevation of the Heat-treating department of the Continental Motor Mfg. Co., Detroit, Mich., Plan of which is shown in Fig. 5. The illustration shows relative location of the oil storage tank, quenching tanks and brine storage tank, as well as the main piping required for the installation.

independently connected with one of the rotary pumps. The proper connections were made to maintain a constant pressure of oil, together with the necessary return piping for the overflow to the tank. A temperature record was obtained by means of smoke charts made on Bristol recording pyrometers. Both the indicating and recording pyrometers were carefully calibrated. The tests were made under actual working conditions and the same kind and quantity of material was heat-treated each day. The material placed in the chamber consisted of fifteen camshafts each weighing 21 pounds. Each camshaft was packed in a 3½-inch steel pipe with the carbonizing material. The total weight of each tube, with shaft and container, was 55 pounds. All the tubes were placed on a frame made of narrow strips of bar steel. This frame held the steel pipes in position and made the total weight in each chamber 950 pounds.

This department is only operated ten hours each day. The material is taken out at 5 P. M. and the furnaces are allowed to cool over night. In the morning the furnace pyrometers register about 700 degrees F. At 6.30 A. M. the burners are lighted and the material is placed in the furnaces which are continued in operation until about 5 P. M. The furnaces reach 1700 degrees about 9 A. M., and they show a uniform heat throughout the working chamber about one hour later. The tests were made under these conditions. The furnaces would have shown higher efficiency if they had been run continuously twenty-four hours every day.

The points noted in making these tests were: Time required for the furnace to reach 1700 degrees; time to obtain a uniform heat in the furnace; evenness of burning and regulation; and amount of oil consumed. The results of the tests follow:

	Fuel Oil	Kerosene
Time tests started	6.30 A.M.	6.30 A.M.
Time required for chamber to reach 1700 degrees F.	9.00 A.M. (2 hr. 30 min.)	8.45 A.M. (2 hr. 15 min.)
Time for chamber to reach uniform heat.	10.15 A.M. (3 hr. 45 min.)	9.45 A.M. (3 hr. 15 min.)
Time test stopped	4.45 P.M.	4.15 P.M.
Fuel consumption required for chamber to reach 1700 degrees F.	26.5 gal.	23 gal.
Additional gallons to reach uniform heats	3.5 gal.	3 gal.
Number gallons for operating furnace 6½ hours after chamber reaches uniform heat	18 gal.	13 gal.
Number gallons per chamber, per hour after uniform heat is reached	2.77 gal.	2 gal.
Number gallons to heat-treat one pound of metal after uniform heat is reached.	0.057 gal.	0.041 gal.
Specific gravity of fuel	41 deg. Baumé	50 deg. Baumé
B.T.U. per pound of fuel	19,890	19,917
Temperature of fuel	65 deg. F.	65 deg. F.

The comparative cost of fuel based on these tests shows that kerosene is from 23 to 25 per cent higher than fuel oil, but the tests also show that there is a saving of thirty minutes' time for the furnace to reach a uniform heat, and a saving of nearly 20 per cent in fuel, by using kerosene. Each burner has a ½-inch oil line and a 1½-inch air line and two burners are connected to each chamber. Power tests show that the blower requires 470 watts for each burner, which is equivalent to 0.63 horsepower.

Modern Types of Gas- and Oil-fired Furnaces. — In the last decade many improvements have been made in furnaces in which steel is heated, to obtain greater accuracy and uniformity in the temperatures. Only a few years ago any temperature that was above the temperature at which steel would harden and below that at which it would become burnt, was considered good enough. In most steels this would cover a range of something like 300 degrees F. Recent investigation, however, has shown that only a few degrees of variation in temperature between these two points makes considerable difference in the hardness, the elastic limit, the reduction of area, or the longevity of steel, as shown by fatigue tests. For instance, to heat steel 50 degrees above the transformation or critical point, or the point at which it should be quenched for hardening, shows a loss of something like 15 per cent in these physical properties; and greater variations show correspondingly greater losses. Thus, while large furnaces formerly had a variation of some 50 to 100 degrees in different parts of the heating chamber, the modern furnace must be so designed that the temperature will not vary more than 10 degrees between any two places, when the furnaces are operated at temperatures that are between 1400 and 1800 degrees F.

The interior construction of the furnaces has, therefore, been changed; appliances for pre-heating the fuel have been devised; gas and oil burners have been improved; automatic heat control instruments have been attached; and heat measuring instruments of various kinds have been brought into use to register and record the temperatures. Other improvements have also been made to reduce the fuel consumption; considerable study has been devoted to this part of furnace designs.

The consumption of fuels has been studied by the Bureau of Mines, at Washington, D. C. One of the things conclusively proved is that a solid wall is better than a hollow wall, especially if the air space is near the outer side of the furnace. The general belief has been that air spaces built into the walls of a furnace would greatly reduce the amount of heat that was dissipated through the walls. The investigation mentioned proved, however, that while heat would travel slowly through air, because it is a poor conductor, the heat would readily leap the air space by radiation, and thus a considerable percentage was lost that could have been saved if the air space had been filled with some solid non-conductor of heat.

These results caused the Industrial Furnace Co., of Detroit, to design and build furnaces in the manner shown in Fig. 7. In these *A* is the cast-iron shell

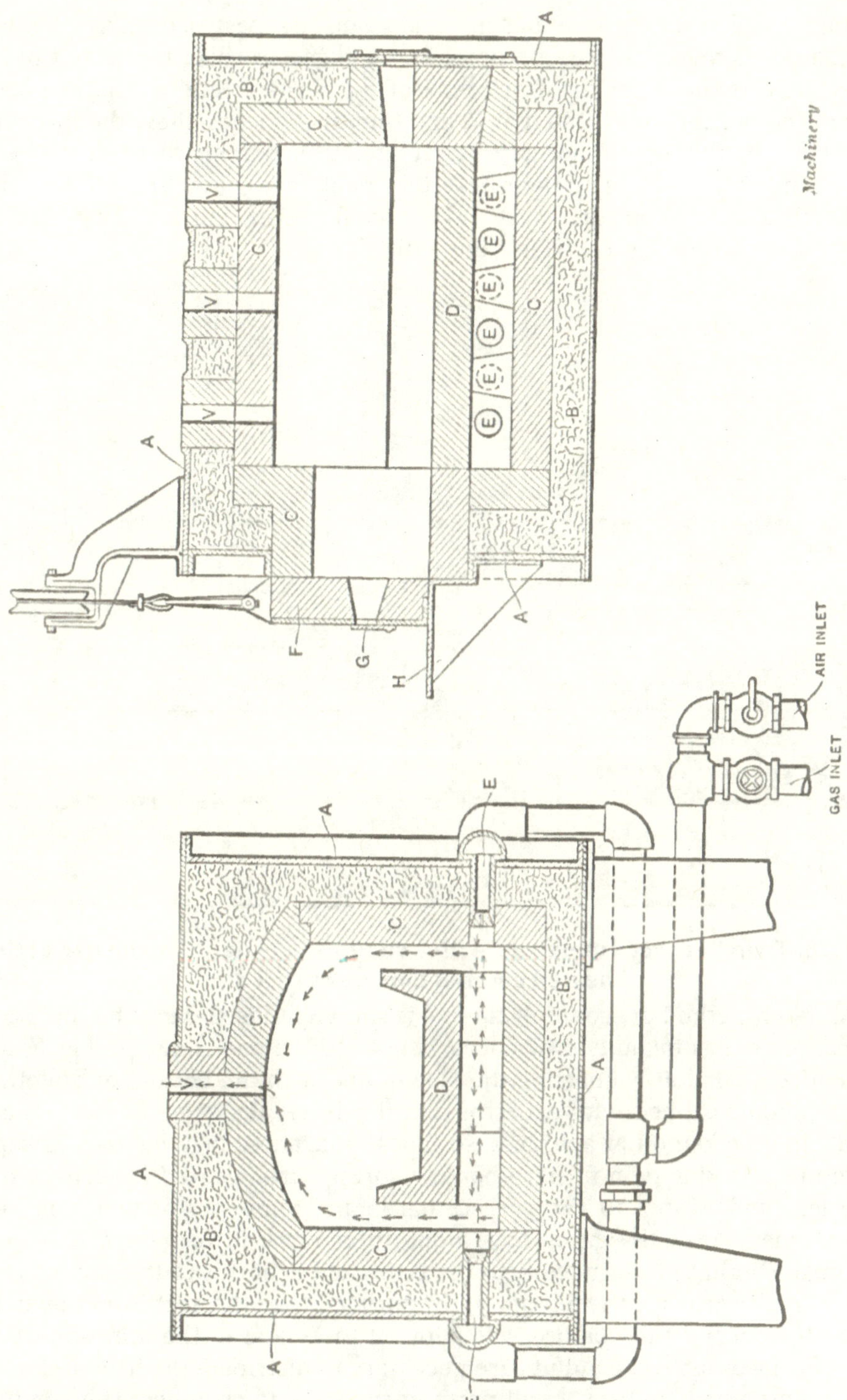

Fig. 7. Heating furnace heavily lined with asbestos to retain the heat

or outer wall of the furnace; *B* is mineral wool, or asbestos, used as a heating insulator; its thickness varies from 2 to 4 inches according to the size of the furnace; *C* is the inner fire-brick wall of the furnace; and *D*, the fire-brick floor. The burners are located at *E*, and the arrow heads show the direction in which the flames and hot gases circulate. After passing through the heating chamber and giving up their heat, the spent gases pass through the vents *V*. At *F* is a sliding door that is typical of such furnaces, and at *G* the peep-hole in the door, while *H* is a shelf in front of the furnace.

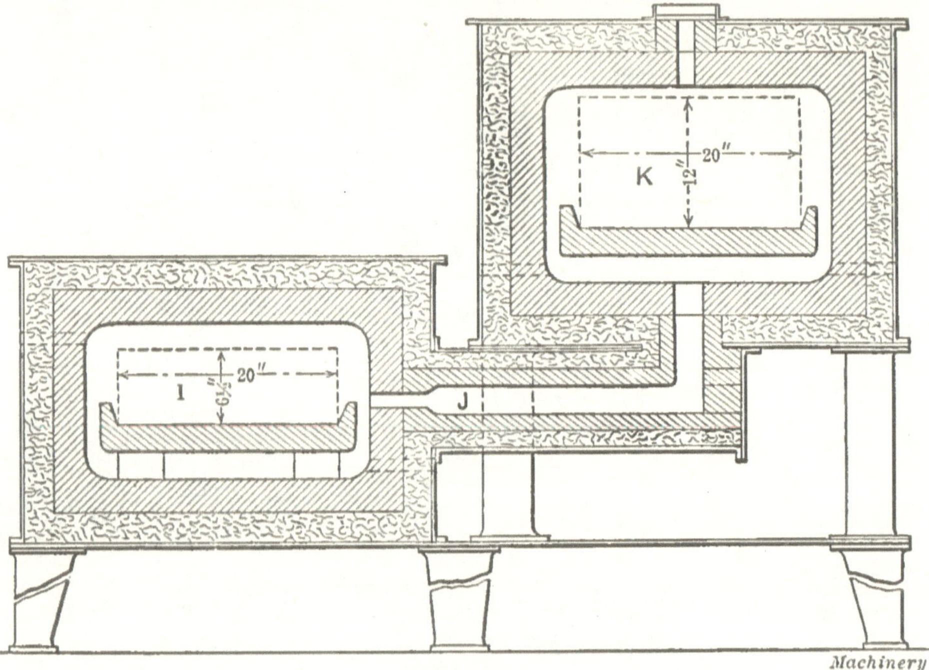

Fig. 8. Twin Furnace connected with a conduit in order to make use of the heated combustion gases twice

Another method of economizing fuel is shown in Fig. 8. Here the Industrial Furnace Co. has taken two furnaces, similar to the one shown in Fig. 7, and provided a conduit connecting the side of one with the bottom of the other. This conduit is lined with asbestos and fire brick, the same as the furnace, and can be taken off at any time, so that the furnaces may be used as separate units. In this twin furnace harrow springs are inserted in furnace *I*, to the left, and heated to the correct hardening temperature which is here maintained. After the gases have done their work in furnace *I*, they pass through conduit *J* to furnace *K*, to the right, and heat this to the correct temperature for drawing the temper of the springs. Thus the heat from the fuel is used the second time before it is allowed to escape to the atmosphere. In long furnaces several conduits are necessary to distribute the heat and make the temperature uniform in all parts of the heating chamber. The conduits should then be provided with dampers, so that the heat can easily be controlled. With only one conduit, however, the temperature within the heating chamber of the tempering furnace can be controlled by opening and closing the vent hole.

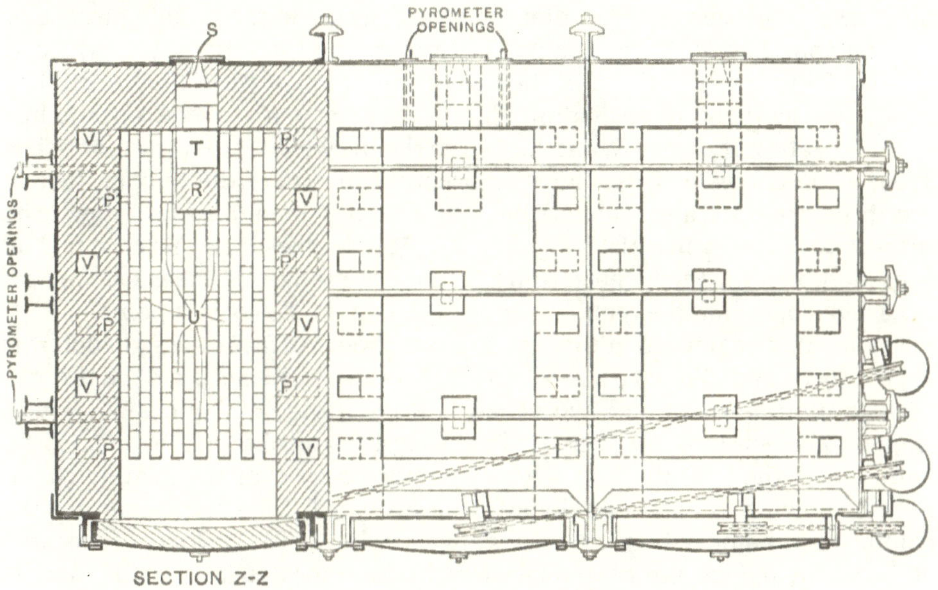

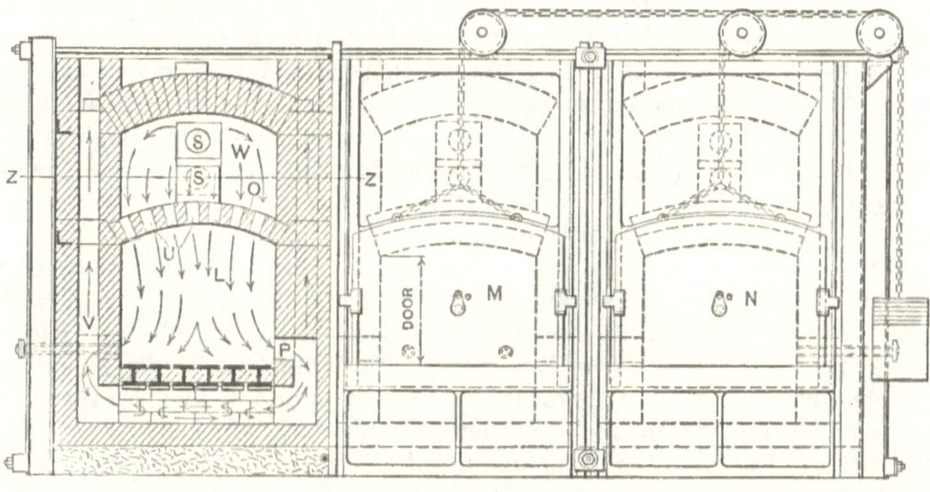

Fig. 9. Over-fired furnace containing three heating chambers arranged for accurate heat control

The same principle has been used by the Garrett-Tilley Furnace Co., of New York, in a three-chambered furnace, each one of which is maintained at a different temperature. These are built in a single unit as shown in Fig. 9; that is, the three different heating chambers are built inside of one furnace shell. In one instance this type of furnace has been used for manufacturing leaf springs. In that case, the heating chamber *L* is used to heat the spring plates to the fabricating heat, which is about 1800 degrees F. When taken from this fire the plates are bent to the correct shape to fit the leaf below, on which they have their bearing. After that they are inserted in the middle furnace or heating chamber *M*, to be heated to the hardening temperature, which is around 1500 degrees F. When heated to this temperature they are taken from furnace *M* and quenched in oil. After that they are inserted in the furnace or heating chamber *N*, and heated to the drawing temperature, which is about 750 degrees F. This allows an accurate control of the temperature in the three separate heating chambers, and each is maintained all day at its respective temperature of 1800, 1500 and 750 degrees F. In this case fuel oil is used for heating the furnace, and pyrometers are used to measure the heat in each oven, so that the temperature can be kept at the correct point.

The heating chambers in this furnace are about six feet in length, and it is especially designed to give a uniform temperature in all parts. On the test run the variation between any two places in each of the heating chambers was shown to be less than 10 degrees. This accuracy was obtained by over-firing and passing the heat through a honey-combed arch over the heating chamber, as shown by the sectional view at in furnace *L*. This arch separates the combustion chamber *W* from the heating chamber *L*. The burners are located at *S*, and the flames enter the furnace at *T* where they strike a baffle plate *R*. This distributes them to both sides of the combustion chamber. After filling the chamber, the hot gases pass through the openings in the honey-combed arch, as shown at *U*, and heat the oven in which the work is placed. The spent gases then leave the heating chamber through ports *P*, pass underneath the floor of the furnace and up through the vents *V*. Thus the top, sides and bottom of the heating chamber are kept at the same temperature throughout its entire length. This over-fired type of furnace has been used in furnaces with a single heating chamber as well as in those that have two and three heating chambers and has proved very successful.

Still another improvement in oil-fired furnaces was recently patented by Walter S. Rockwell of The Rockwell Co., New York. This is shown in Fig. 10. It consists of a pipe coil through which the air is passed and pre-heated before it reaches the burners, where it is mixed with the fuel oil. This pre-heating coil *A* is located in front of the furnace, directly over door *B*, where it receives the heat which comes through the opening in which the work is inserted into the furnace. The plate *C* in front of the coil serves the purpose of protecting the furnace operator from the heat which comes through door opening *B*. One of these furnaces has been used by the Detroit-Timken Axle Co. for some time; it is claimed that the fuel consumption has been reduced by more than twenty-five per cent over that of furnaces that do not pre-heat the air. All or part of the waste gases can be made to pass through the door opening instead of through vents, and thus pre-heat the air to any degree that is desired.

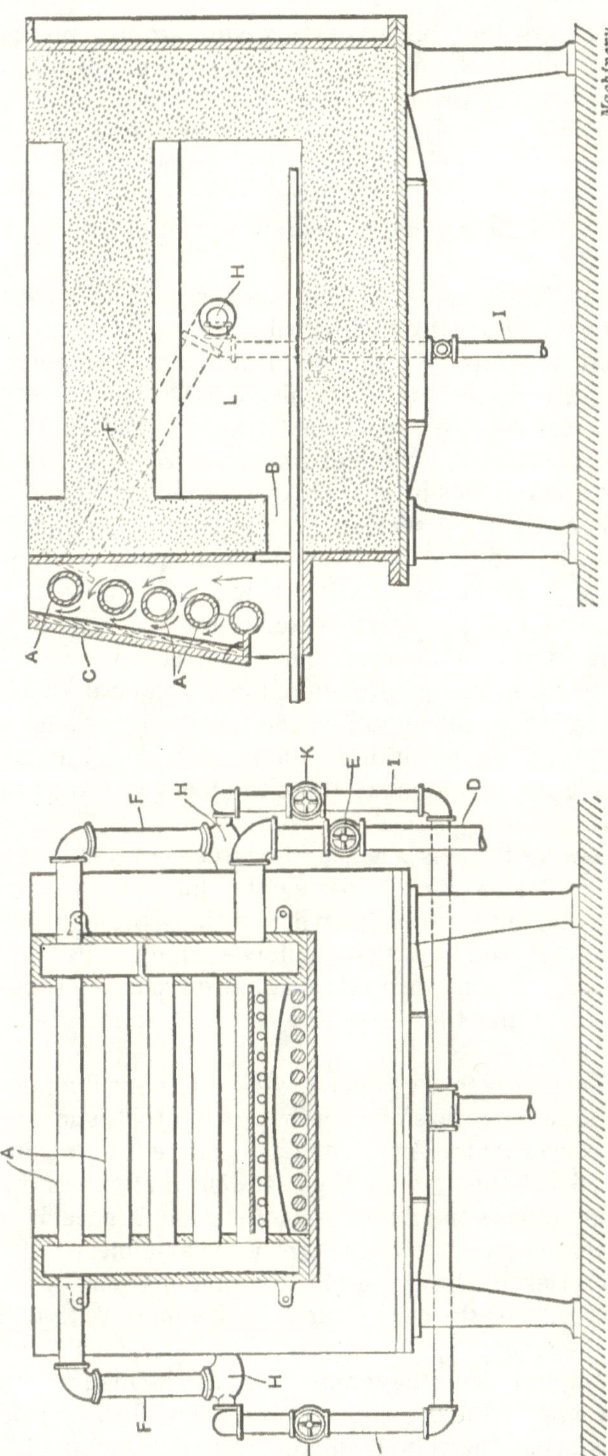

Fig. 10. Recently developed type of oil-burning furnace in which the air is preheated. A pipe through which the air passes before it reaches the burners where it is mixed with the fuel oil, is placed in a preheating chamber in front of the furnace. All or part of the waste gases can be made to pass through the door opening instead of through the vents, and the air can thus be preheated to any degree desired.

The air for combustion enters the coil, under pressure from a pump, through pipe *D*, which is provided with valve *E* to regulate this pressure, so that the air and fuel oil will have the proper mixture. The heated air leaves the coil through pipes *F* and enters burners *H*, where it is mixed with the fuel oil which flows to the burners through pipes *I*, its rate of flow being controlled by valves *K*. The blast of hot gases then passes from burners *H* into heating chamber *L* to raise it to the correct temperature, and out through door *B* to heat the air in coil *A*.

Advantages of Oil and Gas Furnaces. — Oil- and gas-fired furnaces are a great improvement over those that are fired with coal and coke, as with the latter it is impossible to keep the temperature in the furnace at a given point, and much of the heat is lost through the chimney which must be provided to carry away the smoke and gases. A large part of the furnace operator's time is taken up in shoveling in the coal or coke and carrying away the ashes, while the dust and dirt that accumulates from these operations is, to say the least, very disagreeable. Then again the steels heated in such furnaces are more liable to oxidize and scale, and to absorb some of the sulphur or other injurious elements that arise from the combustion. It is a well-known fact that steel absorbs such impurities readily when heated to the high temperatures required for hardening, forging and welding. For these reasons, furnaces using liquid fuels have improved the quality of the metal heated in them, effected a considerable saving in fuel consumption, and saved time by allowing the operator to give more of his attention to the heating of the metal. They have also effected a big improvement in the cleanliness of rooms in which furnaces are located.

When gas and oil furnaces were first installed, a 50-degree variation in the temperature during a day's run or in different parts of the furnace was considered quite good performance, but recent improvements have brought this to a point where a lo-degree variation is all that is allowed in high-class furnaces. This has made it possible to heat-treat steels at more accurate predetermined temperatures, and thus give them greater strengths and resistance to fatigue.

One of the greatest improvements that has been made for controlling the heat in the gas furnace is the temperature control instrument that is manufactured and attached to furnaces by the American Gas Furnace Co. This automatically increases and reduces the amount of gas and air that enters the burners and hence raises and lowers the flame that enters the furnace. It is operated by a mechanism that is attached to the pyrometer. By means of a diaphragm this mechanism raises and lowers a sleeve containing gas and air ports, and thus increases or reduces the size of the port openings. With this instrument the heat inside the furnace can be kept within five degrees of the point at which the instrument is set, and the temperature can be maintained within this narrow limit as long as the gas and air blast keeps flowing into the furnace. The oil-burning furnace must be regulated by an adjustment of

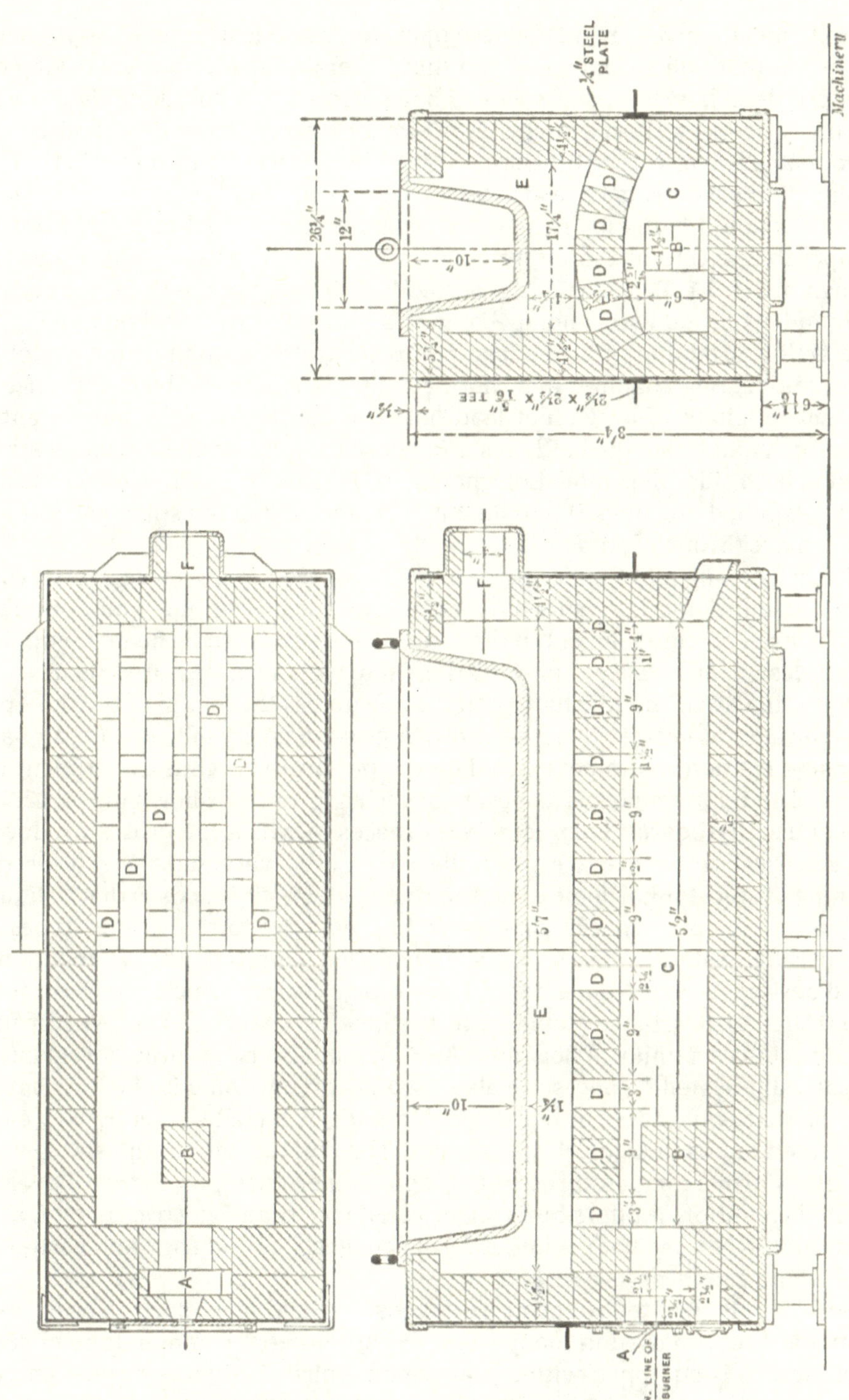

Fig. 11. Furnace for heating liquid bath.

the oil and air valves, by the furnace operator, as no instrument has yet been perfected that will automatically do this. Several individuals are working on this problem, however, and seem to have arrived at a solution. Thus it will probably be but a short time before a similar instrument will be devised for automatically controlling the temperatures in furnaces using oil for fuel.

Many improvements have also been made in oil- and gas-burning furnaces that heat liquid baths for raising the temperature of steel to the hardening temperatures. One of the improvements made in an oil-burning furnace is shown in Fig. 11. This was designed by W. S. Quigley of the Quigley Furnace & Foundry Co. and uses a honey-combed arch similar to that shown in Fig. 9. The arch is used underneath a lead pot to separate the combustion chamber from the heating chamber and more evenly distribute the heat underneath the entire length of a five-foot lead pot. The flames from the burner enter opening A and strike the baffle plate *B* where they are broken up and distributed to both sides of combustion chamber *C*. The hot gases then pass through honey-combed openings *D* into heating chamber *E* and the spent gases leave the furnace through vent *F*.

Salt hardening and tempering bath furnaces are finding more users every day, and oil tempering baths have been used for a long time and doubtless will be used to a great extent in the future. These can also be heated with this same design of furnace. Many of the fluid bath furnaces are constructed without the arch, and it is hardly applicable unless the length of the fluid pot is several times greater than the width. A great majority of lead and salt bath furnaces contain round pots and then the perforated arch is a detriment instead of an improvement. The design of such a furnace is shown in Fig. 12.

Electrically-heated Furnaces. — Furnaces in which the heat is produced by electrical resistance are generally considered very satisfactory for the heating of steel for hardening, but the cost of electricity exceeds that of liquid or gaseous fuels. A special chapter will be devoted to the various types of electrically-heated furnaces that are employed. One type that is commonly used derives its heat from a heavy, low-voltage current which passes through electrodes to resistance elements in the heating chamber. This type of furnace produces a uniform heat and the heat can also be accurately regulated. Electrically-heated furnaces are also used in conjunction with heating baths, the current being transmitted through a bath of metallic salts by two electrodes on opposite sides of the crucible. The conductivity of the salt is very small at normal temperatures, but at high temperatures, when the salt is in a molten condition, it offers but a slight resistance to the electric current, and, therefore, when the bath is hot, it forms an electric conductor and each part of the bath produces its own heat.

Solid Fuels for Steel-heating Furnaces. — Solid fuels, such as coal, coke, charcoal, etc., are used in many cases in hardening. A common type of solid fuel furnace is equipped with a grate upon which the fuel is burned and an arch above the grate which reflects the heat back to the plate that holds the

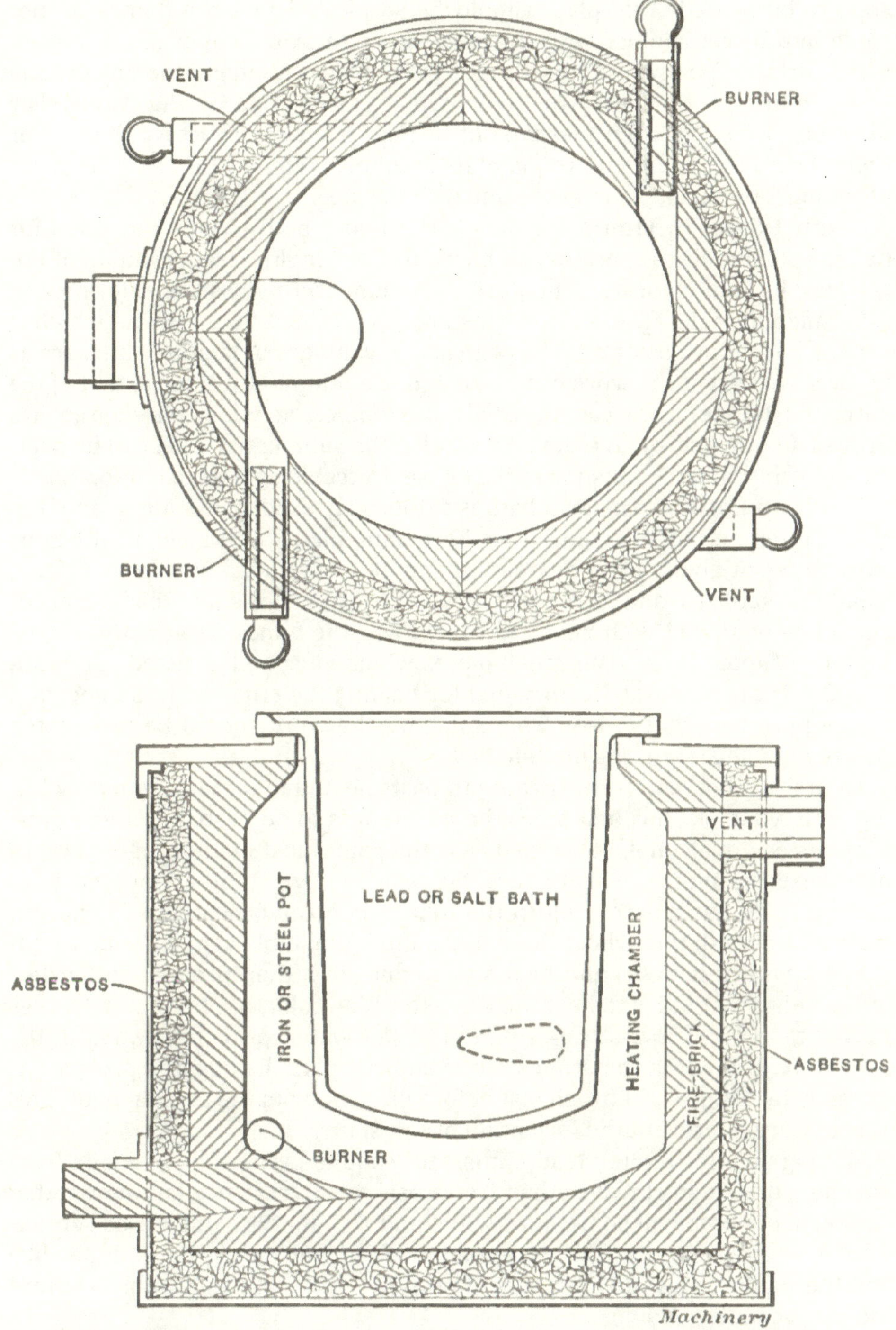

Fig. 12. Simple modern type of furnace for heating lead or salt baths

steel to be heated. This plate should be so located that the flames do not come into direct contact with the steel, so as to oxidize and injure the finished surfaces. To prevent this, the steel is sometimes safeguarded by placing it inside of a clay or cast iron retort which is encircled by the flames. For most purposes, the solid fuel type of furnace is less satisfactory than other types, because it is difficult to maintain a uniform temperature and the gases of combustion are liable to cause injury to the steel.

Heating Steel in Liquid Baths. — The liquid baths commonly used for heating steel tools preparatory to hardening are molten lead, cyanide of potassium, barium chloride, a mixture of barium and potassium chloride and other metallic salts. The molten substance is retained in a crucible which is usually heated by gas or oil. The principal advantages of heating baths are as follows: No part of the work can be heated to a temperature above that of the bath; the temperature can be easily maintained at whatever degree has proved, in practice, to give the best results; the submerged steel can be heated uniformly, and the finished surfaces are protected against oxidation.

The Lead Bath. — The lead bath is extensively used, but is not adapted to the high temperatures required for hardening high-speed steel, as it begins to vaporize at about 1190 degrees F., and, if heated much above that point, rapidly volatilizes and gives off poisonous vapors; hence, lead furnaces should be equipped with hoods to carry away the fumes. Lead baths are especially adapted for heating small pieces which must be hardened in quantities. Gas is the most satisfactory fuel for heating the crucible. It is important to use pure lead that is free from sulphur. The work should be pre-heated before plunging it into the molten lead.

To prevent hot lead from sticking to parts heated in it, mix common whiting with wood alcohol, and paint the part that is to be heated. Water can be used instead of alcohol, but in that case the paint must be thoroughly dry, as otherwise the moisture will cause the lead to "fly." Another method is to make a thick paste according to the following formula: Pulverized charred leather, 1 pound; fine wheat flour, 1½ pound; fine table salt, 2 pounds. Coat the tool with this paste and heat slowly until dry, then proceed to harden. Still another method is to heat the work to a blue color, or about 600 degrees F., and then dip it in a strong solution of salt water, prior to heating in the lead bath. The lead is sometimes removed from parts having fine projections or teeth, by using a stiff brush just before immersing in the cooling bath. This is necessary to prevent the formation of soft spots.

Melting pots for molten lead baths, etc., should, preferably, be made from seamless drawn steel rather than from cast iron. Experience has shown that the seamless pots will sometimes withstand six months' continuous service, whereas cast-iron pots will last, on an average, only a few days, under like conditions. Cast-steel melting pots, if properly made, are as durable as those made of seamless drawn steel.

Cyanide of Potassium Bath. — Many steel hardeners prefer cyanide of potassium to lead, for heating steel cutting tools, dies, etc. When cyanide is used, the parts should be suspended from the side of the crucible by means of wires or wire cloth baskets, to prevent them from sinking to the bottom. Steel will not sink in a lead bath, as lead has a higher specific gravity than steel. Cyanide of potassium should be carefully used, as it is a violent poison. The fumes are very injurious, and the crucible should be enclosed with a hood connecting with a chimney or ventilating shaft. This bath is extensively used for hardening in gun shops, in order to harden parts and at the same time secure ornamental color effects.

Barium-chloride Baths. — A temperature up to 2200 degrees F., and even higher, can be obtained by the barium-chloride bath, and this bath, therefore, is used to some extent for heating high-speed steel for hardening. There are certain disadvantages, however, met with in the use of barium chloride, and for this reason it has been discarded by many manufacturers. This subject is dealt with in greater detail in a following chapter, in connection with electrical hardening furnaces using barium chloride baths. In Fig. 13 is shown a gas-heated furnace with a barium-chloride bath, which has been used by Wheelock, Lovejoy & Co. This furnace is used for heating high-speed steel, the operations in connection with its use being as follows:

The tools to be hardened are first pre-heated, using a small gas furnace. The pre-heating saves time in the barium bath, and is absolutely necessary to avoid checking or cracking the tools, as will be conceded when it is remembered that the temperature of the barium bath is kept at between 2100 and 2200 degrees F. After the tools are pre-heated, they are immersed in the barium bath, being suspended by an iron wire, or, in the case of small parts, in sheet nickel baskets. The reason for using sheet nickel for the baskets is that chloride of barium has a slight dissolving effect on iron, and the exposure of a large area of sheet iron in the bath would eventually destroy the baskets. Nickel is not affected to a perceptible extent, nor is the thin iron wire used to suspend ordinary tools.

The temperature of the barium bath is regulated by a thermoelectric pyrometer. This instrument, shown at the left in Fig. 13, is similar to an ammeter or voltmeter, and the fire end is a thermoelectric couple. The heat of the bath affects the thermo-electric couple and generates a current that deflects the indicator of the indicating instrument to correspond with the temperature. For convenience in operation, the indicating instrument is provided with a double hand, one hand, *A*, being controlled by the temperature of the bath, while the other, *B*, is a marker set by the operator to indicate the temperature which he desires to carry. This marker is made with a disk at the end that covers a hole in the indicating hand when the two coincide, as they do when the temperature has reached the predetermined point. Thus, an operator whose eyes are dazzled by the bright light of the bath does not have to painfully study the graduations to see whether the pointer has reached the correct position, but by glancing at the instrument, he can readily determine when the indicator is directly beneath the marker referred to.

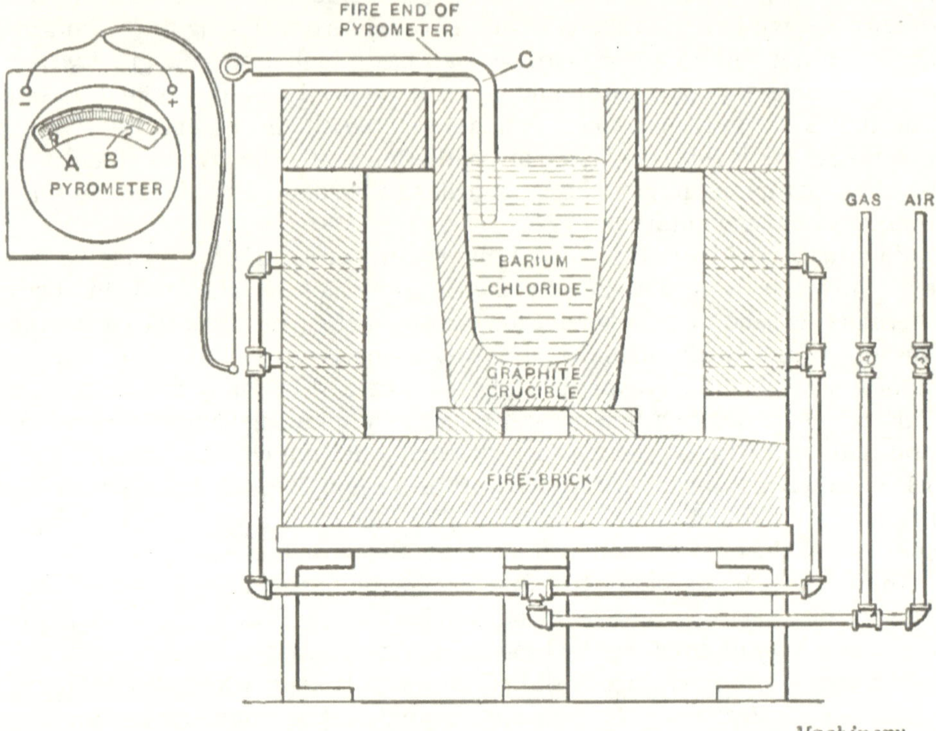

Fig 13. Vertical cross-section of furnace used for heating barium-chloride bath

The immersion of a piece pre-heated to a dull red immediately causes the indicator to drop, the temperature of the bath falling perhaps 30, 40 or even 50 degrees. The fall in temperature is due to absorption of heat by the piece, being the same as the refrigerating effect of a lump of ice thrown into a pot of boiling water, and several minutes may be required to raise the temperature of a large piece to the temperature that is required. For hardening "Blue Chip" steel, a temperature of from 2120 to 2140 degrees F. has been found most suitable. After this temperature is attained, the part is allowed to soak for a few moments, and then is lifted out and dipped into the cooling bath, which consists of cotton-seed oil agitated by compressed air admitted at the bottom. The cotton-seed oil is contained in a large iron barrel surrounded by water in a wooden tub. The part hardened is allowed to remain in the bath until it is quite cold. In practice, the operator hardens a batch and then removes the pieces by means of a wire basket hanging immersed in the oil. It is recommended that milling cutters, end mills, slitting saws, etc., made of "Blue Chip" steel, be used in general, without drawing the temper. However, an oil bath heated by gas and regulated by a thermometer should be provided for tempering such tools as require it.

Chloride of barium is a white transparent salt (Ba Cl_2+2 H_2O) which melts at a temperature of about 1700 degrees, the water of crystallization being driven off at a much lower heat. The salt volatilizes at an extremely high temperature, the loss at the temperature required for heating high-speed steel being negligible. The waste because of volatilization is, say, two pounds from a mass of barium weighing 75 pounds when held at a temperature between 2000 and 2300 degrees for five hours. This property of the chloride of barium bath of standing high temperatures without rapid volatilization is joined with others equally important. The piece heated is protected from the atmosphere during the heating period by the bath, of course, but the protective influence extends still further. A thin coating of barium clings to the piece when it is lifted out for immersion in the cooling bath, thus preventing oxidation. The coating of barium remaining when dipping prevents the coating of burned oil so troublesome to remove, so that on the whole the process probably produces the cleanest work of any bath known. The effect of the barium-chloride bath on the steel is, however, as already mentioned, not altogether satisfactory.

Fig. 13 shows what is considered an improved form of the furnace and crucible used for the chloride of barium bath. The common form of furnace and crucible in use employs a comparatively shallow crucible, which necessitates making a joint between the top of the crucible and the firebrick cover. This gives trouble by loosening and permitting the hot gases to escape around the edge of the crucible. The improved construction utilizes a deeper crucible, the top of which comes flush with the firebrick cover and simplifies the construction. The deep crucible also gives a greater volume of chloride of barium, consequently the refrigerating effect of the pre-heated steel parts, when immersed in the bath, is not so great. This illustration also shows the fire end, C, of the pyrometer immersed in the bath. It has been found advisable to employ crucibles made for steel melting, the ordinary graphite crucible used for brass melting giving trouble by flaking off into the barium.

Barium-chloride Baths used for Carbon Steel. — When a barium-chloride heating bath is used for the lower temperatures required for carbon steels, it is mixed with chloride of potassium. For temperatures between 1400 and 1650 degrees F., use three parts of barium chloride and two parts of chloride of potassium. For higher temperatures, the amount of potassium chloride should be proportionately reduced. When temperatures of 2000 degrees F. and over are required, pure barium chloride must be employed. In all cases, steel heated in barium chloride baths should be pre-heated to 600 or 800 degrees F. before being immersed in the bath. If temperatures below 1075 degrees F. are required, these may be obtained by using equal parts of potassium nitrate and sodium nitrate. This mixture sets at a temperature of 400 degrees F. and is used as a tempering bath, the range of heat obtained with this bath covering practically all the ordinarily used tempering heats.

Value of Pyrometer for Gaging the Heat. — All modern hardening rooms are now provided with pyrometers in order to insure uniformity of the heat in the hardening furnaces, and, consequently, uniform results in the hardened product. The lack of pyrometers, the failure to use pyrometers when provided, the hardening in charcoal furnaces insufficiently heated, and the heating of high-speed steel without pre-heating, are a few of the causes that produce un-uniform results. It seems to be the general opinion among those who do not obtain uniform results when using pyrometers in hardening, that pyrometers are not of much use, that they do not give correct readings, and that as good or better results can be obtained by depending upon a man's experience in this particular work. It is true that many have met with difficulties in the use of pyrometers, but at the same time there is a remedy for this, and that is frequent calibration. While there is no question but that an experienced man's eye is a better judge of the heat in a furnace than an incorrectly calibrated pyrometer, it must be conceded that it is possible to keep the pyrometers in such condition that the readings are a great deal more accurate than any estimate of heat by the eye.

One large firm making lathe and planer cutters for tool-holders has sent out thousands of cutters since the adoption of scientifically correct methods and the use of pyrometers without receiving practically any complaints, which indicates the possibilities of the modern methods. Furthermore, the experience of this firm proves that with the proper heat-treatment any one of the high-grade high-speed steels is entirely satisfactory and that cutters made from any of these best brands cannot be told apart when in use; but the heat-treatment for each, of course, must be suited to that particular steel. In the hardening of high-speed steel, this firm pre-heats all the cutters in a low-heat furnace to a temperature of 1350 degrees F. This heating removes all the internal strains in the metal and puts it in the best possible condition for bringing it quickly to the high heat necessary for high-speed steel. Every cutter is treated in accordance with its sectional area and size, and when placed in the high-heat furnace it remains there for a length of time that has been determined to be correct, for each size, by test. The pyrometers are used both for the low-heat and high-heat furnaces and these pyrometers are tested twice a week. It has been found that this is as long as a pyrometer can be safely employed without checking. The hardening room of this firm is provided with a specially made clock which starts on the pull of a lever by the man running the furnace, the clock having previously been set to indicate the proper length of time required for the cutter being treated. At the end of the predetermined time, the clock rings a bell and stops, and the operator takes out a cutter and puts it into the proper cooling medium.

The experience of this concern indicates that it is necessary to have the high-heat furnace at a temperature above the melting point of the high-speed steel in order to get the best results. This, of course, requires absolute accuracy in the length of time that the work is permitted to remain in the furnace, as a minute too much would ruin the work, and too short a time would not give sufficient heat. Another reason for this high heat in the furnace is that to get the best results from high-speed steel, it is necessary, according to the experience of this firm, to raise the heat from 1350 degrees F. to the highest heat required in a very short period of time. The essentials for proper heat-treatment of high-speed steel are, therefore: A high-grade quality of high-speed steel, pre-heating and rapid rise in the temperature of the steel from the pre-heating temperature to the proper quenching heat in a high-heat furnace, the use of an accurately calibrated pyrometer, and a correct time chart indicating the length of time that each particular piece should remain in the high-heat furnace.

Thermo-electric Pyrometer. — The most commonly used pyrometers for the heat-treatment of steel are of the thermo-electric type. In this type, temperature variations are determined by the measurement of an electric current generated by the action of heat on the junction of two dissimilar metals; that is, when one junction of the thermo-couple has a temperature different from the other, a current is developed and a meter indicates the temperature, the relation between the strength of current and the temperature being constant. The thermocouple and the meter form the essential parts. The two dissimilar metals composing the thermo-couple are connected at one end, which is called the "hot end," and placed in the furnace or heated place, the temperature of which is required. Except at the hot end, the two wires or elements do not touch. The free ends, called the "cold end," are kept away from the heat. When the hot end is heated, the intensity of the current generated depends upon the difference between the temperature of the hot and cold ends. The meter is connected to the cold end and shows the value of the current in degrees Fahrenheit or Centigrade. Some pyrometers of this type may be used, intermittently, for temperatures up to 3000 degrees F.

Resistance Pyrometer. — The variation in electric conductivity due to changes in temperature is the principle upon which the resistance pyrometer is based. This type is very accurate for temperatures below 1600 degrees F., but should not be used continuously for higher temperatures. The maximum temperature is about 2200 degrees F. The thermo-electric type is preferable for indicating high-speed steel hardening temperatures, etc., because the resistance type will not stand exposure to intense heats, except for short periods.

Different Instruments for Measuring Temperatures. — A comparative table of the various types of thermometers and pyrometers in use, explaining the general principles upon which each depends for its working, and stating the limits of temperature between which each type may be used is given herewith. This table is particularly valuable on account of the concise form in which the information it contains has been presented. At the present time pyrometry is becoming an important subject in industrial life, and is not any longer confined to the scientific laboratory only.

Types of Heat Measuring Instruments in General Use

Class	General Characteristics of Action	Type	Range in Degrees Fahrenheit over which they can be used
Expansion	Change in volume or length of a body with temperature.	Gas..............	32 to 1800
		Mercury, Jena glass and nitrogen............	−40 to 900
		Glass and spirit or petrol........	−350 to +100
		Unequal expansion of metal rods............	32 to 900
		Contraction of porcelain.......	32 to 3250
Transpiration and Viscosity	Flow of gases through capillary tubes or small apertures.	The Uehling......	32 to 2900
Thermo-electric	Electromotive force developed by the difference in temperature of two thermo-electric junctions.	Galvanometric....	32 to 2900
		Potentiometric...	32 to 2900
Electric Resistance	Increase in electric resistance of a wire with temperature.	Direct reading on indicator or bridge and galvanometer......	32 to 2200
Radiation	Heat radiated by hot bodies.	Thermo-couple in focus of mirror..	32 to 18,000
		Bolometer........	32 to 18,000
Optical	Change in brightness or in wavelength of the light emitted.	Photometric comparison.........	32 to 3600
		Incandescent filament in telescope..........	32 to 3600
		Nicol with quartz plate and analyzer...........	32 to 3600
Calorimetric	Specific heat of a body raised to a high temperature.	Copper or platinum ball with water vessel....	32 to 2700
Fusion	Unequal fusibility of various metals or earthenware blocks.	Alloys of various fusibilities......	32 to 3600

Calibration of Pyrometers. — Pyrometers should occasionally be compared with a standard pyrometer, or be calibrated in some other way. (The Government Bureau of Standards at Washington will test and calibrate temperature gages for a moderate fee.) The following methods of testing the accuracy of thermoelectric pyrometers are recommended by the Hoskins Mfg. Company: The most satisfactory method is to compare the installed pyrometer with an accurate "check pyrometer" under service conditions. This can be done by placing the thermocouple of the check pyrometer in the same protecting tube with the service thermo-couple, being careful to see that both couples are inserted to the full depth of the tube; then by comparing the readings of the two meters, the accuracy of the service pyrometer can be determined. The check pyrometer should also be calibrated occasionally. For a low-reading type, the calibration may be done by placing the thermo-couple in a heated oil bath and comparing the readings of the meter with those of a high-grade mercury thermometer having a range up to 150 degrees C. (302 degrees F.). The vessel containing the oil should be, approximately, 6 inches deep, and from 4 to 6 inches in diameter, and be heated by a gas burner or an electric hot-plate. The oil should be stirred constantly to insure a uniform temperature throughout. Any oil having a high flash point, or some of the waxes, such as paraffin, beeswax, etc., may be used.

Calibrating by the Melting Point of Copper. — For calibrating pyrometers for temperatures above a red heat, the welded or "hot end" of the thermo-couple should be covered with a tight winding of No. 14 or 16 B. & S. gage, standard melting-point wire. The couple should then be inserted in a tube furnace with the welded end approximately in the center. The furnace should be of the required heat before inserting the couple, and should be kept at a temperature approximately 100 degrees F. higher than the melting point of the calibrating wire. The pointer of the meter will then move up the scale with a gradually decreasing speed until the calibrating wire begins to melt, when the pointer will come to rest. After the wire has melted, the pointer will again move upward. Pure copper wire, under oxidizing conditions, melts at 1065 degrees C. (1949 degrees F.), and pure zinc wire, at 419 degrees C. (786 degrees F.). In order to have a strictly oxidizing atmosphere, an open-end electric furnace should be used for calibrating. With this method of calibrating, care should be taken not to have the furnace temperature too far above the melting point of the calibrating wire, because the pointer will move so rapidly and the melting will be of such short duration that the temporary pause of the pointer may not be observed.

Calibrating by the Freezing Point of Melted Salts. — A very satisfactory way of calibrating pyrometers is by using the "freezing points" of melted salts. Pure common salt (NaCl) is melted in a pure graphite crucible. When the salt has been raised to a temperature of 100 to 200 degrees F. above its melting point, the bare welded end of the thermo-couple is inserted to a depth of 2 or 3 inches. The crucible is then removed from the furnace and allowed to cool. The pointer on the meter will drop gradually until the salt begins to freeze or solidify; then the pointer will stop until the salt is frozen. The freezing point of pure salt is taken at 800 degrees C. (1472 deg. F.). After calibrating and before being further used, the couple end should be washed in hot water to remove all traces of the salt, as otherwise the couple will deteriorate rapidly, especially when heated considerably above the melting point of salt in an open furnace. When calibrating pyrometers, care should be taken that the zero setting of the meter agrees with the cold end of the couple, which is always kept away from the heat and generally at the temperature of the outside air. The following table gives the latest available data by the Bureau of Standards on certain substances which may be used for calibrating pyrometers.

Water boils at ....100 deg. C. (212 deg. F.)
Tin freezes at ....232 deg. C. (450 deg. F.)
Zinc freezes at ....419 deg. C. (786 deg. F.)
Common salt freezes at ..... 800 deg. C. (1472 deg. F.)
Copper, in oxidizing atmosphere, freezes at 1065 deg. C. (1949 deg. F.)

Correcting the Thermo-couple. — If the pyrometer gives too high a reading, the correction is made by increasing the adjusting resistance, and if the reading is too low, the resistance is decreased. The adjusting resistance of a Hoskins pyrometer is No. 26 B. & S. gage wire, which has a low temperature coefficient and is wound on a composition spool located within the handle of the thermo-couple. In changing this resistance, the connection should always be carefully soldered and the different turns insulated from one another. This changing of the resistance should not be attempted by an inexperienced person.

Chapter Three - Quenching and Tempering

While there has been a great deal published regarding the hardening and tempering of steel and the furnaces used, there is but little detailed information available regarding the baths used for these operations. The time has long since passed when each hardener had his own carefully guarded secrets regarding the composition of quenching baths. On the other hand, the time has also passed when it was a common belief that to harden a piece of steel it was only necessary to cool it off more or less rapidly in almost any kind of cooling medium; it has been found that the cooling mediums for hardening and the heating mediums or baths for tempering do, after all, play quite an important part both as regards economy and the efficiency of the tools treated.

It is the intention in this chapter to outline the methods which have proved most successful, and to give the compositions of baths which from long experience in connection with hardening operations have proved to give the best all around results and to be the most economical; in many cases they have not been the cheapest in initial cost, but nevertheless are most economical, because of the better results obtained with the tools treated and the greater length of time that the baths could be used before deteriorating. A description of such receptacles — cooling tanks and tempering furnaces — for the treatment of steel as have proved to be the best for all around purposes has also been included.

Characteristics of Quenching Baths. — No matter what the composition of a quenching bath, to insure uniform hardening the temperature of the bath must be kept constant, so that successive pieces of steel or tools quenched will be acted upon by baths of the same heat. The necessity of a uniform temperature for a quenching bath will be readily understood by reference to ordinary water for a cooling bath; everyone having any knowledge of the subject knows that a tool quenched in such a bath at room temperature will come out much harder than if quenched when the water is at the boiling point. In fact, it is well known that one way of partially annealing steel is by plunging it at a red heat into hot water. The same difference in hardness will result when using any quenching bath at different temperatures, and hence no dependable results can be obtained unless means are taken for keeping these baths at a uniform heat.

When using quenching baths of different composition the tools quenched will vary in hardness. This is due mainly to the difference in heat-dissipating power of the different baths. Thus a tool hardened at the same temperature in water and brine will come out harder when quenched in brine; the greater the conductivity of the bath the quicker the cooling. The general opinion, today, is that the composition of a quenching bath is of small importance as long as the bath cools the pieces rapidly. Those who have made a study of the subject have found different opinions regarding the same quenching bath by different users, and a good many quenching fluids have been condemned owing to improper heating which in turn was due, in many cases, to improperly built furnaces. As an example may be cited an oven furnace with which the user once had trouble. Owing to faulty construction of this furnace, more air was let into the heating chamber of the furnace than could be taken care of by the fuel oil; after having condemned first the steel and then the quenching bath, and then trying one quenching bath after another with the same results, it was suggested that the "heating" did not look just right, and an expert was called in to find out what the trouble was. After much experimenting with the burners and the furnace itself good results were finally obtained. The difficulty seemed to be that the oxygen of the air attacked the steel and formed oxide of iron on the surface of the tools, which consequently had a soft scale on the outside.

Those who are skeptical as to there being any difference in the effect on steel of cooling baths of different composition will readily admit that it is advantageous to use baths free from oxygen and from ingredients that tend to oxidize. Quenching baths should be uniform; good tool steels of high carbon are very sensitive to differences in both water and oils. Water for hardening tool steel should be soft; entirely different and very unsatisfactory results will be obtained when using hard water. While different quenching oils show less difference in the results obtained, vegetable and animal oils will give somewhat different degrees of hardness depending upon the sources from which they are obtained. One cannot be too careful in the selection of water, as it is likely to contain many impurities. If it contains greasy matters, it may not harden steel at all, whereas if it contains certain acids, it will be likely to make the tools quenched in it brittle and even crack them.

Quenching Baths for Specific Purposes. — Clear cold water is commonly employed for ordinary carbon steel, and brine is sometimes substituted to increase the degree of hardness. Sperm and lard oil baths are used for hardening springs, and raw linseed oil is excellent for cutters and other small tools. The effect of a bath upon steel depends upon its composition, temperature and volume. The bath should be amply large to dissipate the heat rapidly, and the temperature should be kept about constant, so that successive pieces will be cooled at the same rate. Greater hardness is obtained from quenching in salt brine and less in oil, than is obtained by the use of water. This is due to the difference in the heat-dissipating qualities of these substances. If thin pieces are plunged into brine, there is danger of cracking, owing to the suddenness of the cooling.

A quenching solution of 3 per cent sulphuric acid and 97 per cent of water will make hardened carbon steel tools come out of the quenching bath bright and clean. This bath is sometimes used for drills and reamers which are not to be polished in the flutes after hardening. Another method of cleaning drills and similar tools after hardening is to pickle them in a solution of 1 part hydrochloric acid and 9 parts of water. Still another method is to use a heating bath consisting of 2 parts barium chloride and 3 parts potassium chloride. This method is satisfactory for reamers and tools which are not to be polished in the flutes after hardening. A small quantity of sal-ammoniac added to an oil-bath also has a tendency to make the tools come out clean from the bath.

That the temperature of the hardening bath has a great deal to do with the hardness obtained has been proved by actual tests. In certain experiments a bar quenched at 41 degrees F. showed a scleroscopic hardness of 101. A piece from the same bar quenched at 75 degrees F. had a hardness of 96, while, when the temperature of the water was raised to 124 degrees F., the bar was decidedly soft, having a hardness of only 83. The higher the temperature of the quenching water, the more nearly does its effect approach that of oil, and if boiling water is used for quenching, it will have an effect even more gentle than that of oil; in fact, it will leave the steel nearly soft. With oil baths, the temperature changes have little effect upon the degree of hardness. Parts of irregular shape are sometimes quenched in a water bath that has been warmed somewhat to prevent sudden cooling and cracking. A water bath having one or two inches of oil on top is sometimes employed to advantage for tools made of high-carbon steel, as the oil through which the work first passes reduces the sudden action of the water.

Irregularly shaped parts should be immersed so that the heaviest or thickest section enters the bath first. After immersion, the part to be hardened should be agitated in the bath; the agitation reduces the tendency of the formation of a vapor coating on certain surfaces, and a more uniform rate of cooling is obtained. The work should never be dropped to the bottom of the bath until quite cool. High-speed steel is cooled for hardening either by means of an air blast or an oil bath. Both fresh and salt water are also used, although, as a general rule, water should not be used for high-speed steel. Various oils, such as cotton-seed, linseed, lard, whale oil, kerosene, etc., are also employed; many prefer cotton-seed oil. Linseed has the objection of becoming gummy, and lard oil has a tendency to become rancid. Whale oil or fish oil gives satisfactory results, but has offensive odors, although this can be overcome by the addition of about three per cent of heavy "tempering" oil.

List of Quenching Baths. — (1) Water — soft — preferably distilled; good tool steel should require no mixture added to pure water. (2) Salt added to water; will produce a harder "scale" than if quenched in plain water. (3) Sea (salt) water — the keenest natural water for hardening. (4) Water as under (1), containing soap. (5) Sweet milk. (6) Mercury. (7) Carbonate of lime. (8) Wax. (9) Tallow. (10) Air — mostly used for high-speed steel; mere exposure, however, is in many cases and on many steels not sufficient to produce hardness and an air blast is necessary, as this furnishes cool air in rapid motion. (11) Oils such as cotton-seed, linseed, whale, fish, lard, lard and paraffine mixed, special quenching oils, etc. Milk, mercury, carbonate of lime, wax and tallow are generally used for special purposes only.

The following list of oils and names of firms supplying them is given for the sake of convenience. The firms mentioned are reliable and their oils have been thoroughly tried out in comparison with other makes and have proved to be superior; opinions may, of course, differ in this respect and no doubt there are many oils that have not been tried that may be as good.

Cotton-seed oil — Union Oil Co., Providence, R. I.; Underhay Oil Co., Boston, Mass.

Linseed oil — Spencer Kellogg & Sons, Inc., Buffalo, N. Y.

Whale oil — no difference found between two different kinds.

Fish oil — only one kind tried.

Lard oil — W. B. Bleecker, Albany, N. Y.; E. F. Houghton & Co., Pittsburg, Pa.

Paraffine oil — Underhay Oil Co., Boston, Mass.

Special quenching oil — E. F. Houghton & Co., Pittsburg, Pa. Very good and cheap. While this may possibly deteriorate somewhat faster than some of the others mentioned it will prove very economical.

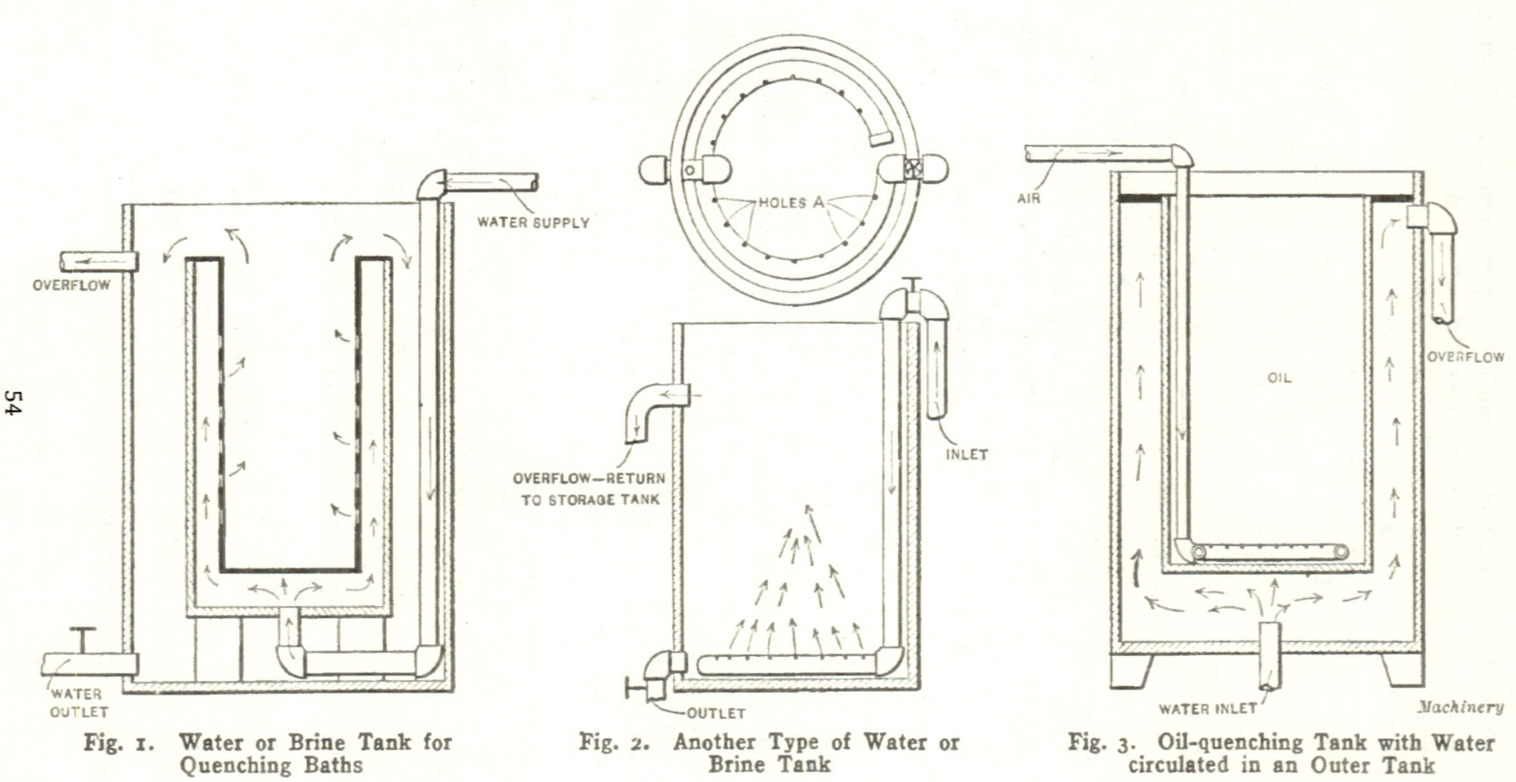

Fig. 1. Water or Brine Tank for Quenching Baths

Fig. 2. Another Type of Water or Brine Tank

Fig. 3. Oil-quenching Tank with Water circulated in an Outer Tank

The order of the intensity with which various cooling baths will harden steel of about 0.90 to 1.00 per cent carbon is as follows: Mercury, carbonate of lime, pure water, water containing soap, sweet milk, different kinds of oils, tallow and wax. In all cases, except possibly the oils, tallow and wax, it must be remembered that the tools become harder as the temperature of the bath becomes lower.

Receptacles and Tanks used in Quenching. — The main point to be considered in a quenching bath is, as mentioned, to keep it at a uniform temperature so that successive pieces quenched will be subjected to the same heat. The next consideration is to keep the bath agitated, so that it will not be of different temperatures in different places; if thoroughly agitated and kept in motion, as is the case with the bath shown in Fig. 1, it is not even necessary to keep the pieces in motion in the bath, as steam will not be likely to form around the pieces quenched. Experience has proved that if a piece is held still in a thoroughly agitated bath, it will come out much straighter than if it has been moved around in an unagitated bath. This is an important consideration, especially when hardening long pieces. It is, besides, no easy matter to keep heavy and long pieces in motion unless it be done by mechanical means.

In Fig. 1 is shown a water or brine tank for quenching baths. Water is forced by a pump or other means through the supply pipe into the intermediate space between the outer and inner tank. From the intermediate space it is forced into the inner tank through holes as indicated. The water returns to the storage tank by overflowing from the inner tank into the outer one and then through the overflow pipe as indicated. In Fig. 2 is shown another water or brine tank of a more common type. In this case the water or brine is pumped from the storage tank and continuously returned to it. If the storage tank contains a large volume of water, there is no need of a special means for cooling. Otherwise, arrangements must be made for cooling the water after it has passed through the tank. The bath is agitated by the force with which the water is pumped into it. The holes at *A* are drilled on an angle, so as to throw the water toward the center of the tank. In Fig. 3 is shown an oil quenching tank in which water is circulated in an outer surrounding tank for keeping the oil bath cool. Air is forced into the oil bath to keep it agitated.

Fig. 4 shows a water and oil tank combined. The oil is kept cool by a coil passing through it in which water is circulated, which later passes into the water tank. The water and oil bath in this case is not agitated.

Fig. 5 shows the ordinary type of quenching tank cooled by water forced through a coil of pipe. This can be used for either oil, water or brine. Fig. 6 shows a similar type of quenching tank, but with two coils of pipe. Water flows through one of these and steam through the other. By this means it is possible to keep the bath at a constant temperature.

Tempering. — The object of tempering is to reduce the brittleness in hardened steel, and to remove the internal strains caused by the sudden cooling in the quenching bath. The tempering process consists in heating the

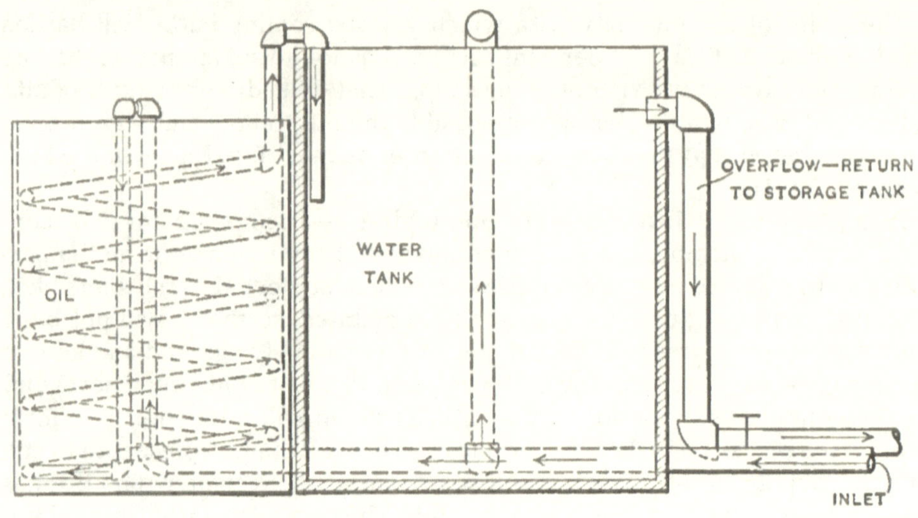

Fig. 4. Water and oil tank combined

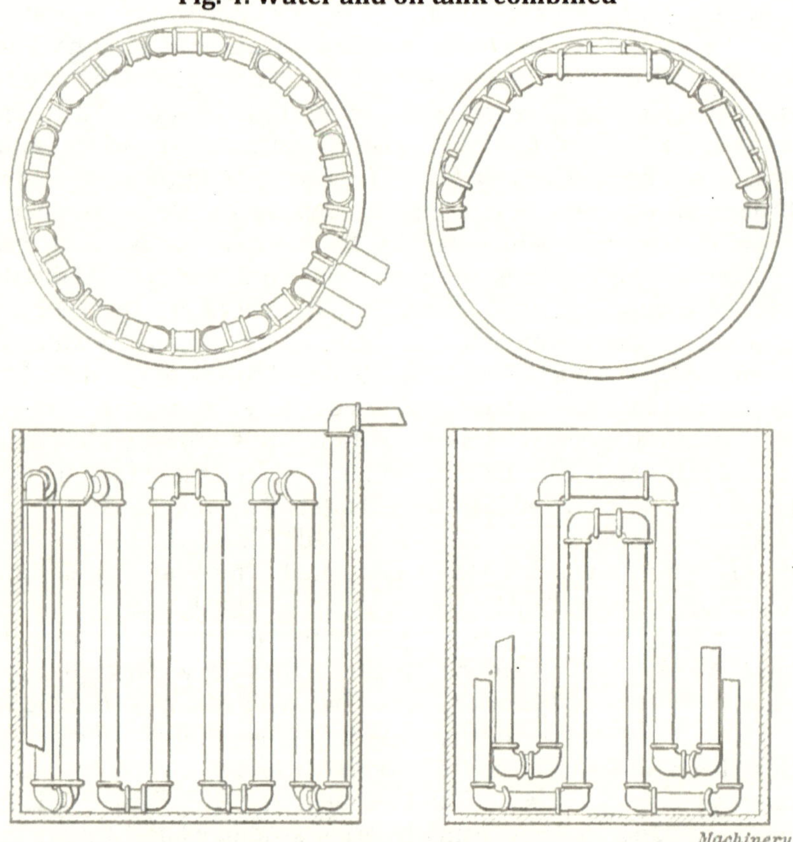

Fig. 5. (Left) Ordinary type of quenching tank
Fig. 6. (Right) Oil-quenching tank with water and steam coils

piece of work, by various means, to a certain temperature, and permitting it to cool off gradually. The degree of heat to which the tool to be tempered is heated determines the degree of toughness — and unfortunately also the degree of softness — it has attained; the higher the tempering heat, the less brittle, and also the less hard, will the tool be.

Tempering by the Color Method. — When steel is tempered by the color method, the tempering heat is gaged by the colors formed on the surface as the heat increases. First the surface is brightened to reveal the color changes, and then the steel is heated either by placing it upon a piece of red-hot metal, a gas-heated plate or in any other available way. As the temper increases, various colors appear on the brightened surface. First there is a faint yellow which blends into straw, then light brown, dark brown, purple, blue and dark blue, with various intermediate shades. The temperatures corresponding to the different colors and shades are given in the table on temperatures and colors for tempering. Turning and planing tools, chisels, etc., are commonly tempered by first heating the cutting end to a cherry-red, and then quenching the part to be hardened. When the tool is removed from the bath, the heat remaining in the unquenched part raises the temperature of the cooled cutting end until the desired color (which will show on a brightened surface) is obtained, after which the entire tool is quenched. The foregoing methods are convenient, especially when only a few tools are to be treated, but the color method of gaging temperatures is not dependable, as the color is affected, to some extent, by the composition of the metal.

Colors for Tempering

Color	Degrees F.	Degrees C.
Very pale yellow	430	221
Light yellow	440	227
Pale straw-yellow	450	232
Straw-yellow	460	238
Deep straw-yellow	470	243
Dark yellow	480	249
Yellow-brown	490	254
Brown-yellow	500	260
Spotted red-brown	510	266
Brown-purple	520	271
Light purple	530	277
Full purple	540	282
Dark purple	550	288
Full blue	560	293
Dark blue	570	299

The modern method of tempering, especially in quantity, is to heat the hardened parts to the required temperature in a bath of molten lead, heated oil or other liquids; the parts are then removed from the bath and quenched. The bath method makes it possible to heat the work uniformly, and to a given temperature within close limits.

Baths used for Tempering. — As the object of tempering is simply to reduce the brittleness and to remove internal strains caused by sudden cooling in quenching, the composition of a tempering bath is of little importance compared with that of a quenching bath when considering the effect upon the pieces treated. Aside from the operator's convenience and possible bad effects upon his health, the different baths used for this operation must be considered with regard to initial cost, lasting quality, effects on finish, etc.

While oil is the most widely used medium for tempering tools in quantities, other means and methods are employed, especially by those who have tools in small quantities to temper, when the expense of installing and running an oil tempering furnace would not be warranted. Of these methods we first find the one already described, used by the old-style tool hardener, of only partly cooling the tool when quenching it, then quickly withdrawing it, polishing off the working surface, and then letting the heat which remains in the tool produce the required temper as judged by the color.

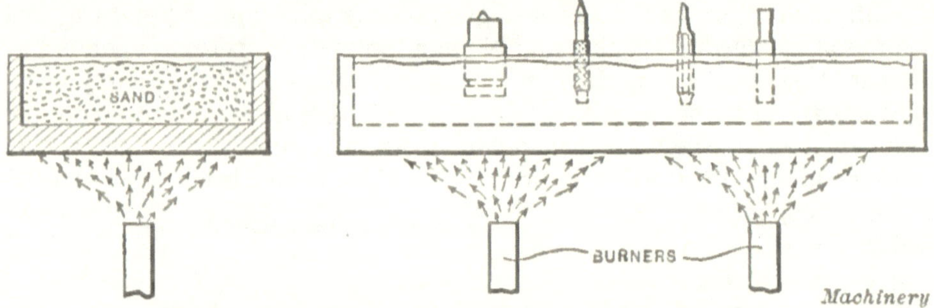

Fig. 7. Arrangement used for Sand Tempering

The sand bath is another frequently used medium for tempering, the sand being deposited on an iron plate and heated; by the use of this method a piece to be tempered can be given different tempers throughout its length, as, for example, rivet hole punches; these are placed endwise — bottom down — in the sand about two-thirds projecting outside the sand into the air (see Fig. 7). It is readily seen that the nearer the bottom of the sand bath, the higher the heat, and the punch so placed, when tempered right, will have the bottom soft — a deep dark blue — the neck a very dark straw, and the working part of the punch on top a light straw color; thus there is a gradual increase in hardness from the bottom up. Pieces so drawn must previously have been polished, and the temper is judged by the color. When the pieces have attained the right color they are, of course, cooled off, generally in water or oil. A plate without sand similarly heated can also be used, but it is not as satisfactory.

A plate arranged as shown in Fig. 8 will be found very convenient when drawing small, round pieces. The pieces are rolled on the inclined plate which is heated as indicated. The length of time the work is in contact with the plate can be regulated by adjusting the amount of the incline, as well as the location of the "stop." This arrangement can also be used for such work as punches, etc., in which case the plate, of course, should stand level and not in an inclined position.

Another frequently used tempering medium is hot air, the temper in this case also being judged by the color. For this method of tempering special furnaces should be employed in order to get uniform results. This method is used more especially for small and light work in quantities and where the color has to be bright and clear. While all of these methods have the advantage of enabling one to actually see the temper given to tools treated, the oil tempering bath is the one mostly used owing to its economy.

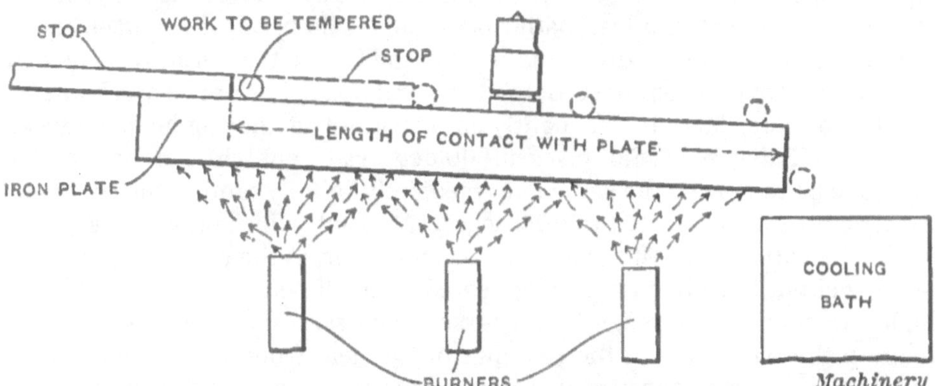

Fig. 8. Tempering Arrangement utilizing an Inclined Plate on which the Objects roll down

The two main points to be considered when using an oil tempering furnace are: first, to have the heat uniform throughout (not hotter where the burners or flames are in contact with the walls of the furnace); and second, to leave the pieces to be tempered in the oil long enough to have attained the heat of the oil throughout when taken out. The first point can be taken care of, as far as possible, by proper construction of the furnace; the second can best be taken care of by immersing the pieces to be tempered in the oil before starting to heat, and letting the pieces remain in the oil and be heated with it to the temperature required. In such a case, one should, of course, have more than one furnace, or else after each operation take the hot oil out and refill the tank with cold. The method described is very much better than the one frequently used of immersing the pieces in a bath which already has the required temperature and then letting them remain long enough to attain the heat of the bath throughout, as a furnace yet has to be designed which will maintain a uniform heat for even as short a time as is required for this operation. Furthermore, it is not necessary that a piece to be tempered be held in the bath a certain length of time at the required temperature; the temperature desired need only be maintained long enough to insure that the piece has been evenly heated throughout.

There are tempering oils on the market claimed to have a flash test of 750 degrees, but it is doubtful if they ever have been found to stand this test. Heavy black cylinder oil has been found to stand a flash test of 725 degrees. Therefore, when tempering to high heats, or, rather, when tempering to higher heats than the flash point of any tempering oils (650 to 700 degrees F.) some other tempering fluid than oil must be used. Lead is the one usually employed. As it is impracticable when using lead to let the pieces to be tempered be heated up with the lead, they must be immersed at the predetermined temperature and kept there until heated evenly throughout to the same temperature as the lead. It is claimed by many that it is easier to maintain a uniform heat in a lead bath than in an oil bath, but it has been found that, owing to the lead not circulating as readily, the temperature may vary considerably in different parts of the bath, and hence it is not very reliable.

Salt is another medium frequently employed for tempering heats between 800 and 875 degrees F. Salt fuses at 800 degrees F., but when immersing the pieces to be tempered the salt will immediately solidify around the cold pieces. When these are heated to 800 degrees, the salt will melt and the pieces should be withdrawn. This is not reliable, however, as the pieces, especially if large, will not have had time to be heated through before the salt melts. If a higher temper is required, it is, of course, only necessary to let the pieces remain in the bath and get the readings of the heat from a pyrometer. In all these methods, it is questionable if it is good practice to suddenly immerse cold pieces to be drawn into baths of such high temperatures. When a lower temper is required, and an oil tempering bath or furnace is not available, alloys of lead and tin can be used for as low heats as 400 degrees and of lead and antimony for 500 degrees. However, this involves the inconvenience of keeping a large number of different alloys on hand, if it is desired to vary the temper heats. The following table for different alloys which melt at the temperatures given was compiled by Mr. O. M. Becker.

Lead	Tin	Melting Temperature, Degrees F.	Lead	Tin	Melting Temperature, Degrees F.
14	8	420	24	8	480
15	8	430	28	8	490
16	8	440	38	8	510
17	8	450	60	8	530
18.5	8	460	96	8	550
20	8	470	200	8	560

These alloys should be carefully made and then run into small strips of about ½ inch square, being afterward remelted. The melting pot should be carefully heated by gas, if possible, and the metal should only be heated to such a point that the insertion of the tool causes it to set round the steel.

When such is the case the steel becomes equally heated as the metal melts, and it can be allowed to remain some time in the metal before taking out and quenching. Another plan is to lay the tools on the cold alloy and allow them to remain until it melts, thus permitting the steel to gradually warm through, and possibly giving better results in the hands of some men. Usually, however, the plunging of the tools into the molten metal is the method adopted, and provided time is given to allow for the absorption of heat to a sufficient depth, this gives good, tough tempering with the requisite hardness.

The oils for tempering baths specified below are given for the sake of convenience only; the statements are based upon the findings of thorough experiments. There may, of course, be many other oils just as good that have not been tried.

(1) Walter A. Wood, Boston, Mass., XXX tempering oil; as cheap in initial expense as any; good lasting qualities.

(2) Strong, Carlisle & Hammond Co., Cleveland, Ohio, Frankfort tempering oil.

(3) Fish oil, cotton-seed or linseed oil may also be used; in many cases these are mixed with high fire and flash test mineral oils.

The analysis and test results of oils when new (not used) as compared with those of oils which have been used such a length of time as to render them practically valueless will be found interesting.

Properties of Oil	W. A. Wood Tempering Oil		Lard and Paraffine Oil Mixed (Half and Half) Used for Quenching	
	New	Old (Thick)	New	Old (Thick)
Flash point	550	475	400	380
Fire test	625	550	475	450
Mineral oil, per cent	94	30	25	10
Saponifiable oil, per cent	6	70	75	90
Specific gravity	0.920	0.950	0.912	0.925

Houghton tempering oil: flash point, 595 degrees; fire test, 685 degrees; specific gravity, 0.900.

Frankfort tempering oil: fire test, 670 degrees.

Frankfort quenching oil: fire test, 500 degrees.

Paraffine oil (Underhay): fire test, 450 degrees; specific gravity, 0.912.

Lard oil (Bleecker): fire test, not determined; specific gravity, 0.920.

The great difference in tests and analysis between new and used oils should be noted; oils used constantly at high heats will gradually lose the "mineral" part of the oil, the more so the higher the heat used. A tempering bath can therefore be prolonged in life by adding to it now and then new mineral oil. To lengthen the life of the bath high heats should be avoided as much as possible.

Practical Tempers for Carbon Steel Tools. — This list of tempers was determined by practical shop tests of the tools mentioned. A record was kept of the number of pieces machined before the tool required sharpening or renewal, and the most satisfactory temper adopted. A thermometer was used to determine degrees of heat. Mutton tallow was used for the bath.

Degrees F.	**Class of Tool**
495 to 500	Taps ½ inch or over, for use on automatic screw machines.
490 to 495	Taps ½ inch or over, for use on screw machines where they pass through the work.
495 to 500	Nut taps ½ inch and under.
515 to 520	Taps ¼ inch and under, for use on automatic screw machines.
525 to S30	Thread dies to cut thread close to shoulder.
500 to 510	Thread dies for general work.
495	Thread dies for tool steel or steel tube.
440 to 445	Circular thread chaser for use on lathes.
525 to 540	Dies for bolt threader threading to shoulder.
460 to 470	Thread rolling dies.
430 to 435	Hollow mills (solid type) for roughing on automatic screw machine work.
450 to 455	Hollow mills (solid type) for use on the drill press.
485	Knurls.
450	Twist drills for hard service.
450	Centering tool for automatic screw machine.
430	Forming tools for automatic screw machine.
430 to 435	Cut-off tools for automatic screw machine.
440 to 450	Profile cutters for milling machine.
430	Formed milling cutters.
435 to 440	Milling cutters.
430 to 440	Reamers.
460	Counterbores and countersinks.
440 to 450	Fly-cutters for use on the drill press.
480	Cutters for tube or pipe-cutting machine.
430 to 440	Dies for heading bicycle spokes.
430	Punches for heading bicycle spokes.
430	Backer blocks for spoke drawing dies.
400	Drawing dies for bicycle spokes.
460 and 520	Snaps for pneumatic hammers — Harden full length, temper to 460 degrees, then bring point to 520 degrees.

Tempering Furnaces. — In tempering furnaces the only really important consideration is to insure that the furnace is so built as to heat the bath uniformly throughout. It is doubtful if there can be found a tempering furnace on the market that will fill this requirement entirely, although many give good results in general. It is never safe, however, to let any tools being tempered rest against the bottom or sides of the tank, as no matter how scientifically the furnace may be built these parts are, in most cases, hotter than the fluid itself. It is, of course, just as important not to let the thermometer rest against any of these parts in order to insure correct readings. After the pieces tempered are taken out of the oil bath, they should immediately be dipped in a tank of caustic soda (not registering over 8 or 9), and after that in a tank of hot water. This will remove all oil which might adhere to the tools.

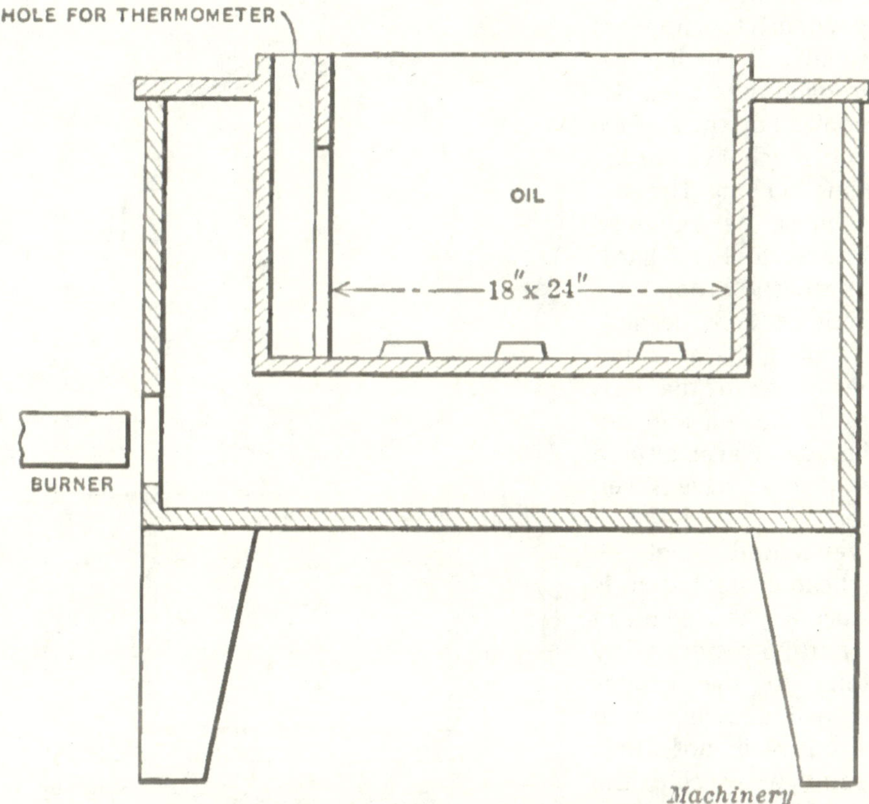

Fig. 9. Ordinary Type of Tempering Furnace

Fig. 9 shows an ordinary type of tempering furnace. In this the flame does not strike the walls of the tank directly. The tools to be tempered are laid in a basket which is immersed in the oil. In Fig. lo is shown a tempering furnace in which means are provided for preventing the tools to be tempered from coming in contact with the walls or bottom of the furnace proper. The basket holding the tools is immersed in the inner perforated oil tank. This same arrangement can, of course, be applied to the furnace shown in Fig. 9.

Defects in Hardening. — Uneven heating is the cause of most of the defects in hardening. Cracks of a circular form, from the corners or edges of a tool, indicate uneven heating in hardening. Cracks of a vertical nature and dark-colored fissures indicate that the steel has been burned and should be put on the scrap heap. Tools which have hard and soft places have been either unevenly heated, unevenly cooled, or "soaked," a term used to indicate prolonged heating. A tool not thoroughly moved about in the hardening fluid will show hard and soft places, and have a tendency to crack. Tools which are hardened by simply dropping them to the bottom of the tank, sometimes have soft places, owing to contact with the floor or sides of the tank. They should be thoroughly quenched before dropping. When a tool appears soft and will not harden, it probably has been decarbonized on the surface by too much heat or by soaking too long. The surface must be removed before the tool will harden properly. Tools are sometimes soft because the cooling bath is not large enough for the tools being hardened, and becomes too warm after a few pieces have been quenched.

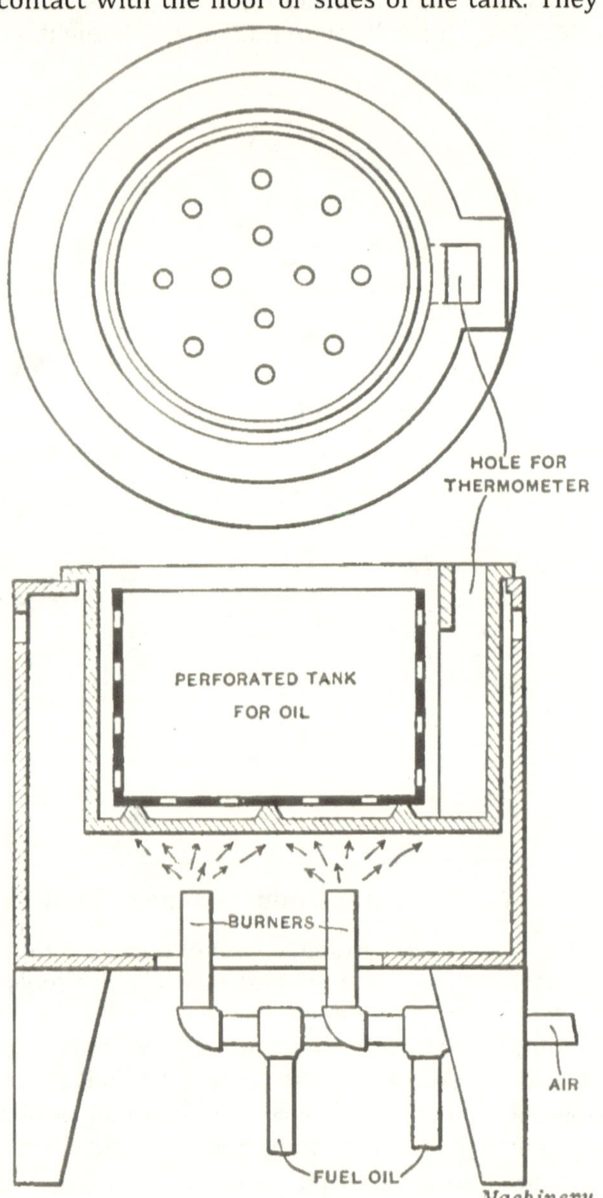

Fig. 10. Special Tempering Furnace with Perforated Oil Tank

Overheated Steel. — Overheated steel that is not actually burned can be *partly* restored by heating to the proper heat, and allowing it to cool slowly in hot ashes or sand; when cold, the steel is hardened again at the proper hardening heat. Tools treated in this way are not as good as when treated at the proper heat throughout, but they are partially restored, and if the overheating originally took place in forging, the risk of cracking in hardening will be lessened by adopting the process mentioned. Care should be taken that the tuyere of the forge is well covered when heating tool steel; a tool coming in direct contact with the air blast will become surface burned, show soft places in hardening and wear badly in use.

Scale on Hardened Steel. — The formation of scale on the surface of hardened steel is due to the contact of oxygen with the heated steel; hence, to prevent scale, the heated steel must not be exposed to the action of the air. When using an oven heating furnace, the flame should be so regulated that it is not visible in the heating chamber. The heated steel should be exposed to the air as little as possible, when transferring it from the furnace to the quenching bath. An old method of preventing scale and retaining a fine finish on dies used in jewelry manufacture, small taps, etc., is as follows: Fill the die impression with powdered boracic acid and place near the fire until the acid melts; then add a little more acid to insure covering all the surfaces. The die is then hardened in the usual way. If the boracic acid does not come entirely off in the quenching bath, immerse the work in boiling water. Dies hardened by this method are said to be as durable as those heated without the acid.

Annealing Steel. — The purpose of annealing is not only to soften steel for machining, but to remove all strains incident to rolling or hammering. A common method of annealing is to pack the steel in a cast-iron box containing some material, such as powdered charcoal, charred bone, charred leather, slaked lime, sand, fireclay, etc. The box and its contents are then heated in a furnace to the proper temperature, for a length of time depending upon the size of the steel. After heating, the box and its contents should be allowed to cool at a rate slow enough to prevent any hardening. It is essential, when annealing, to exclude the air as completely as possible while the steel is hot, to prevent the outside of the steel from becoming oxidized.

The temperature required for annealing should be slightly above the critical point, which varies for different steels. Low-carbon steel should be annealed at about 1650 degrees F., and high-carbon steel at between 1400 and 1500 degrees F. This temperature should be maintained just long enough to heat the entire piece evenly throughout. Care should be taken not to heat the steel much above the decalescence or hardening point. When steel is heated above this temperature, the grain assumes a definite size for that particular temperature, the coarseness increasing with an increase of temperature. Moreover, if steel that has been heated above the critical point is cooled slowly, the coarseness of the grain corresponds to the coarseness at the maximum temperature; hence, the grain of annealed steel is coarser, the higher the temperature to which it is heated above the critical point.

If only a small piece of steel or a single tool is to be annealed, this can be done by building up a firebrick box in an ordinary blacksmith's fire, placing the tool in it, covering over the top, then heating the whole, covering with coke and leaving it to cool over night. Another quick method is to heat the steel to a red heat, bury it in dry sand, sawdust, lime or hot ashes, and allow it to cool. Quick annealing can also be partially effected by heating the piece to a dull black-red and plunging it into hot water. This method is not to be recommended.

Chapter Four - Heat-Treatment of High-Speed Steel

Hardening High-speed Steel. — High-speed steel must be heated to a much higher temperature for hardening than carbon steel. A temperature of from 1400 to 1600 degrees F. is sufficient for carbon steel; high-speed steel requires from 1800 to 2200 degrees F. The usual method of hardening a high-speed steel tool, such as a turning or planing tool, is to heat the cutting end slowly to a temperature of about 1800 degrees F., and then more rapidly to about 2200 degrees F., or until the end is at a dazzling white heat and shows signs of melting down. The tool point is then cooled either by plunging it in a bath of oil (such as linseed or cotton-seed) or by placing the end in a blast of dry air. When an oil quenching bath is used, its temperature is varied from the room temperature to 350 degrees F., according to the steel used. The exact treatment varies for different steels and it is advisable to follow the directions given by the steel makers. High-speed steel parts that would be injured by a temperature high enough to melt the edges are hardened by heating slowly to as high a degree as possible and then cooling, as described. Formerly, the air blast was recommended by most steel makers, but oil is now extensively used. Care should be taken to quench the heated steel rapidly after removing from the source of heat. The barium-chloride bath has been used quite extensively for heating machine-finished, high-speed steel tools preparatory to hardening. The barium-chloride forms a thin coating on the steel, which i^s thus protected from oxidation while being transferred from the heating bath to the cooling bath. Tests have demonstrated, however, that barium-chloride baths have certain disadvantages for heating high-speed steel preparatory to hardening, because if the steel is heated to the required temperature, the surface of the tool is softened to some extent. These tests indicate that whenever this salt is used as a heating bath, the temperature should not be raised above 2050 degrees F. When about 0.010 inch is ground from the cutting edges of the tools, the objectionable influence of heating in barium chloride may be negligible. The influence of barium chloride on steel in hardening is treated completely in the chapter on "Metallic Salt-bath Electric Hardening Furnaces."

The oil used may depend on how hard the tool is desired. Suppose it is required "dead" hard on the cutting edges; this is as hard as it is possible to make it for machine shop use and still retain sufficient toughness. To obtain this result, after fusing the point or cutting edges, quench quickly in thin lard oil, or, for extreme hardness, quench the tool in kerosene oil, when about the maximum hardness of this steel may be obtained. In using oils, especially kerosene, great care should be taken, or the oil may flame and burn the operator. The oil tanks used for hardening should be constructed preferably of galvanized iron, fitted with close-fitting covers, and provided with a screen a few inches from the bottom.

With regard to the hardening and tempering of specially formed tools of high-speed steel, such as milling and gear cutters, twist drills, taps, threading dies, reamers, and other tools that do not permit of being ground to shape after hardening, and where any melting or fusing of the cutting edges must be prevented, the method of hardening is as follows:

A specially arranged muffle furnace heated either by gas or oil is employed, consisting of two chambers lined with fireclay, the gas and air entering through a series of burners at the back of the furnace, and so under control that a temperature up to 2200 degrees F. may be steadily maintained in the lower chamber, while the upper chamber is kept at a much lower temperature. Before placing the cutters in the furnace it is advisable to fill up the hole and keyways with common fireclay to protect them. The cutters are first placed upon the top of the furnace until they are warmed through, after which they are placed in the upper chamber, and thoroughly and uniformly heated to a temperature of about 1500 degrees F., or, say, a medium red heat, when they are transferred into the lower chamber and allowed to remain therein until the cutter attains the same heat as the furnace itself, *viz.*, about 2200 degrees F. and the cutting edges reach a bright yellow heat, having the appearance of a glazed or greasy surface. The cutter should then be withdrawn while the edges are sharp and uninjured, and revolved before an air blast or quenched in oil until the red heat has passed away, and then while the cutter is still warm — that is, *just* permitting of its being handled — it should be plunged into a bath of tallow at about 200 degrees F. and the temperature of the tallow bath then raised to about 520 degrees F., on the attainment of which the cutter should be immediately withdrawn and plunged in cold oil.

Of course there are various other ways of tempering, a good method being by means of a specially arranged gas-and-air stove in which the articles to be tempered are placed, and the stove then heated up to a temperature of from 500 degrees F. to 600 degrees F., when the gas is shut off and the furnace with its contents allowed to slowly cool down.

Very satisfactory results in hardening high-speed steel tools, such as cutters, drills, etc., have been obtained by the following method: First pre-heat in an oven-type gas furnace to from 1300 to 1500 degrees F.; then transfer the steel to another gas furnace having a temperature varying from about 2000 to 2200 degrees F.; when the steel has attained this temperature, quench in a metallic salt bath having a temperature varying from 600 to 1200 degrees F., depending upon the kind of high-speed steel used. The piece to be hardened should be stirred vigorously in the bath until it has obtained the temperature of the bath; then it is cooled, preferably in the air, and requires no further tempering; or it may be put directly into the tempering oil, which should be at a temperature anywhere between 100 and 600 degrees F. The tempering bath is then gradually raised to the heat required for tempering. The salt bath for quenching should consist of calcium chloride, sodium chloride and potassium ferro-cyanide, in proportions depending upon the required heat. Various kinds of steel require different temperatures for the metallic salt bath. After the temper of the tool has been drawn in the oil, the work is dipped in a tank of caustic soda, and then in hot water. This will remove all oil which might adhere to the tools, and is a method that applies to all tools after being tempered.

The method used in the United States Navy yards for hardening high-speed steel cutting tools is as follows: One oil-heated furnace is maintained at a temperature of from 1600 to 1700 degrees F. and another furnace at from 2400 to 2450 degrees F. The cutting ends of the tools are pre-heated in the low-temperature furnace and then transferred to the high-temperature furnace. After the tools are removed from the high-heat furnace they are cooled by dipping the ends into oil. The oil is agitated by compressed air and is cooled and maintained at as even a temperature as possible. The tools are cooled in the oil until they just show a full black, when they are removed and placed on a cooling table.

The Taylor-White Process. — This process of hardening high-speed steel is, in brief, as follows: The first method, commonly known as the "high-heat treatment," is effected by heating the tool slowly to 1500 degrees F., and then rapidly from that temperature to just below the melting point, after which the tool is quickly cooled below 1550 degrees. At this point, the cooling is continued either fast or slow to the temperature of the air. It is important to avoid any increase of temperature during the cooling period. The second or "low-heat treatment" (tempering) consists in reheating a tool which has had the high-heat treatment to a temperature somewhere between 700 and 1240 degrees F., preferably in a lead bath, for a period of five minutes. The tool is then cooled to the temperature of the air either rapidly or slowly.

Tempering High-speed Steel. — Heavy high-speed steel tools having well-supported cutting edges (such as planing or turning tools) are commonly used after hardening and grinding, without tempering. If high-speed steel tools are comparatively weak, they are often toughened by tempering to suit the particular service required. This is sometimes referred to as "letting down" the hardness. The steel is heated in lead, oil or in a forge fire. A method recommended by several steel makers is to cover the steel with clean dry sand and heat to the required temperature, as indicated, preferably, by a pyrometer. The sand is contained in a metal pan and is heated by a gas or oil burner. A general idea of tempering temperatures for high-speed steel may be obtained from the following figures: Milling cutters, 400 degrees F.; threading dies and taps, 490 degrees F.; drills and reamers, 440 degrees F. for large sizes, and 460 degrees F. for small sizes. All heavy turning and planing tools are left untempered.

Annealing High-speed Steel. — Accurate annealing is of much value in bringing the high-speed steel to a state of molecular uniformity, thereby removing internal strains that may have arisen; at the same time, annealing renders the steel sufficiently soft to enable it to be machined into any desired form for turning tools, milling cutters, drills, taps, threading dies, etc. Further advantage also results from careful annealing by minimizing risks of cracking when the steel has to be reheated for hardening. In cases of intricately-shaped milling tools having sharp square bottom recesses, fine edges or delicate projections, and on which unequal expansion and contraction are liable to operate suddenly, annealing has a very beneficial effect toward reducing cracking to a minimum. Increased ductility is also imparted by annealing and this is especially requisite in tools that have to encounter sudden shocks due to intermittent cutting, such as planing and slotting tools, or others suddenly meeting projections or irregularities on the work operated on. The annealing of high-speed steel is best carried out in muffle furnaces designed for heating by radiation only, a temperature of 1400 degrees F. being maintained from twelve to eighteen hours according to the section of the bars of steel dealt with.

A number of other methods are also used for annealing high-speed steel. The following method is recommended by one of the largest high-speed tool steel manufacturers in America. Particular attention is called to the temperatures to which the steel to be annealed is to be heated, the time necessary, and also, that powdered charcoal is given first, it having the preference over fine air-dried lime or powdered mica.

"In annealing high-speed steel, use an iron box or pipe of sufficient size to allow at least one-half inch of packing between the pieces of steel to be annealed and the sides of the box or pipe. (We call attention here to the fact that it is not necessary that each piece of steel to be annealed be kept separate from every other piece, but only that the steel be prevented from touching the sides of the annealing pipe or box.) Pack carefully with powdered charcoal, fine dry lime or mica. Cover with cap, which should be air-tight, but if it is not, then lute on with fireclay. Heat slowly to a full red heat, about 1475 or 1500 degrees F., and hold at this heat from two to eight hours, depending on the size of the pieces to be annealed. A piece of 2 by 1 by 8 inches requires about three hours' time. Cool as slowly as possible, and do not expose to the air until cold. A good way is to allow the box or pipe to remain in the furnace until cold."

A method almost identical with the one just described, is given in somewhat greater detail by a writer in Machinery. This method consists of placing the tool in a wrought-iron or cast-iron tube, having space large enough in circumference and length to accommodate the work and plenty of packing material. One end of the tube can be threaded for a cast-iron cap, which can be screwed on and off as desired; the other end can be permanently fixed on the tube, as it is generally only necessary to use the one end. Have a number of 1/8-inch holes drilled in the tube, and also a few 3/16-inch holes in the end caps. The idea of the holes in the tube is to procure a vent for letting off the gas which is generated when heating the packing material. The end holes can be used for a number of test wires, which can be withdrawn as the heating progresses, and by this means the operator will be able to ascertain the proper heat desired, which should be a bright orange or a trifle higher, according to the nature of the steel. When the desired heat is reached, regulate the blast sufficiently to hold the muffle and its contents at this heat for a period long enough to allow the heat to thoroughly penetrate the steel. Then the muffle with its contents may be buried in dry slaked lime or sawdust and ashes, and allowed to cool down slowly. If a furnace is used for the heating, allow the muffle to stay in until furnace and all cools down, when the steel will be found quite easy to work in the machine. Respecting this "dead heating" in the furnace, unless the steel is properly packed in the muffle in order to exclude the oxygen blown into the furnace by the blast, the steel is apt to oxidize and the carbon content thereby becomes lowered, resulting in over-annealed steel. The packing material used may be any one of several kinds now commonly used for other work, but charred leather gives the best results, and dry, fine charcoal of a good, clean quality is effective.

Experiments relating to the Annealing of High-speed Steel. — A series of experiments was recently made to determine the temperature to which high-speed steel should be heated for annealing. It was found that when this steel was heated to below 1250 degrees F. and slowly cooled, as in annealing, it retained the original hardness and brittleness imparted to the steel in forging. When heated to between 1250 and 1450 degrees F., the Brinell test indicated that the steel was soft, but impact tests proved that the steel still retained its original brittleness. However, when heated to between 1475 and 1525 degrees F. the steel became very soft, it had a beautiful fine-grained fracture, and all of the initial brittleness had entirely disappeared.

In carrying these tests further, to 1600, 1750 and 1850 degrees F., it was found that the steel became very soft, but there was a gradual increase in brittleness and in the size of the grain, until at 1850 degrees F. the steel became again as brittle as unannealed steel; the fracture at this temperature was dull, dry and lifeless, and showed marked decarbonization. Dried air-slaked lime was used as a packing medium in making these tests. The steel was packed in tubes both ends of which were afterward provided with air-tight caps. The decarbonization that took place was probably due to the oxygen in the air that had filled the intervening spaces between each minute particle of lime, before it was packed in the tube, attacking the carbon of the steel; this decarbonization would not have taken place if powdered charcoal had been used. The latter would have supplied all the carbon necessary to combine with any oxygen present in the tubes.

An annealing chart, taken by a Bristol recording pyrometer, showing the temperature of one of the annealing furnaces in which a well-known grade of high-speed steel is annealed by the manufacturer, is shown in Fig. 1. The method, which is carried on by this manufacturer day after day, is to first pack the bars to be annealed in ten-inch diameter wrought-iron pipes, about fourteen feet long, the packing medium being pulverized charcoal. Then both ends of the pipes are sealed air-tight with fireclay. The annealing furnaces are fired with coal and are brought up to 1500 degrees F. at 7 A. M. At this time the large furnace doors are opened and from four to six of the ten-inch pipes, previously packed with steel and sealed, are rolled into the furnace. The doors are then closed and the furnace is continuously fired until 5 130 P. M., the temperature being kept as near to 1500 degrees F. as possible. The chart, which shows two days' work, will indicate how well this temperature has been maintained. At 5:30 P. M. firing is discontinued, all holes that might permit the influx of air are closed, and the pipes are permitted to cool down slowly with the furnace. It will be seen, by again referring to the chart, that there is a gradual drop in temperature from the time firing is discontinued until the pipes are taken from the furnace the following morning preparatory to beginning another day's work.

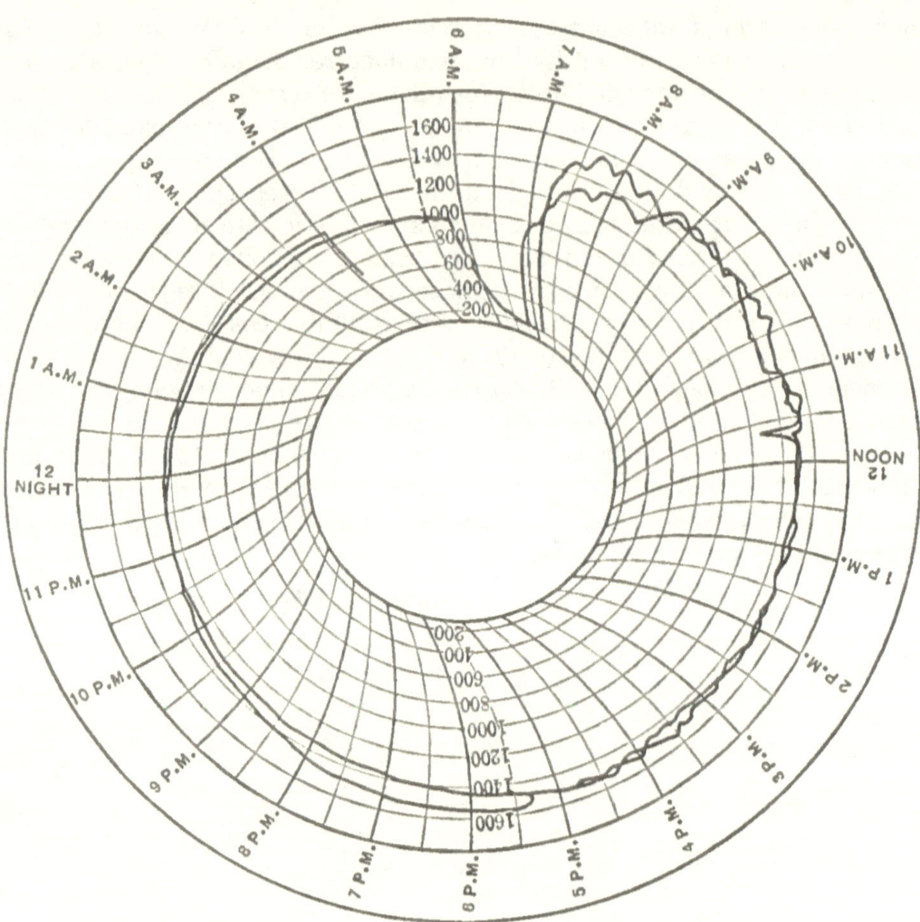

Fig. 1. Annealing Chart from a Recording Pyrometer showing the Temperature of a Furnace used for annealing High-speed Steel

Kind of Steel	Method of Cutting off Unannealed Bars	Forging Heat*	Directions for Annealing			
			Packing and Packing Material	Time Required for Heating	Temperature Required	Manner of Cooling
Burgess No. 4 and No. 5	Cut off hot	Even cherry red	3 parts sand, 1 part lime	2 to 3 hours	Bright red	Very slowly
Ark	Cut off hot	Canary color	....	....	....	....
Midvale Special Self-hardening	Nick on wheel, break cold	Dull orange	1 part lime, 1 part charcoal	....	Light cherry red	Slowly
Capital High-speed	....	Full bright red	Heat in air-tight box lined with fire-brick	1 to 4 hours	1650° F.	Very slowly
Heller's Alloy High-speed	....	Bright yellow	Powdered charcoal; air-tight box	....	Bright yellow (2000° F.)	Slowly
Blue Chip	....	Cherry red	Lime; air-tight box	....	Bright red or dark yellow	Very slowly
Allen's Air Hardening	....	Bright red	Lime; air-tight box	....	Bright red	Very slowly
Bethlehem Self-hardening	Cut off hot	Between bright cherry red and full yellow	Heat and bury in lime	....	Bright cherry red	Slowly
Rex High-speed	Cut off hot	Bright red	Charcoal, lime or sand; air-tight box	2 to 8 hours	Full red	Very slowly
Novo	Cut off hot	High lemon color	Lime; air-tight box	12 to 18 hours; soaking heat	Full red	Very slowly
Bohler's Styrian High-speed	Nick hot, break cold	Bright yellow	Charcoal; air-tight box	4 to 6 hours	1830° F.	Slowly
McInnes Extra High-speed	....	Full red	Ashes or lime; air-tight box	....	Cherry red	Slowly

[*] If common forge is used for heating, have a deep fire, plenty of coke and light blast. In general, an oil, gas or coke furnace is preferred.

Kind of Steel	Hardening Heat, Cooling Medium, Temper and Grinding					
	Lathe and Planer Tools, etc.			Taps, Milling Cutters, etc.		
	Temperature	Cooling Medium	Condition of Wheel used for Grinding	Temperature	Cooling Medium	Temper Required
Burgess No. 4 and No. 5	White welding heat on point	Cold air blast or fish oil†	Dry wheel (grind slowly)	....	....	....
Ark	Fusing heat on point	Cold air blast or thin oil†	Very wet wheel	Light yellow heat	Oil†	....
Midvale Special Self-hardening	White heat	Cold air blast	....	Dark cherry heat; then heat in white hot lead	Oil, cool until color disappears	....
Capital High-speed	White heat	Cold air blast	Wet sandstone	Full white heat	Cold air blast	No temper
Heller's Alloy High-speed	Bright yellow heat	Air blast or fish oil†	Wet stone	Bright yellow heat*	Air blast†	Temper seldom required
Blue Chip	Clear white heat on point	Air blast or oil	Dry wheel	White heat just below fusing	Oil	Straw color
Allen's Air Hardening	White heat on point	Cold air blast or water at 150° F.	Wet stone	....	....	....
Bethlehem Self-hardening	White heat on point	Dry, cold air blast	....	As hot as possible without fusing	Oil	....
Rex High-speed	Fusing heat on point	Cold air blast (cool until tool can be held in the hand)†	Wet or dry	As hot as possible without fusing	Oil†	Temper if necessary
Novo	Fusing heat on point	Cold air blast or running lard or fish oil†	Wet wheel (grind slowly)	Red heat; then white heat in furnace	Thin fish or cottonseed oil†	According to use of tool
Bohler's Styrian High-speed	White heat (not fusing)	Cold air blast	....	White heat, just below fusing	Fish oil	....
McInnes Extra High-speed	Fusing heat on point	Air blast or fish oil	Dry stone	Light cherry red heat	Air blast or fish oil	....

[*] Do not use gas fire.

[†] Steel must not be dipped in water while hot.

The chart also indicates that the temperature of the annealed steel, when taken from the furnace, is about 1000 degrees F. This temperature is several hundred degrees below the critical point, or recalescence point of high-speed steel, this point being at about 1350 degrees F., so that the annealed bars can be taken from the pipes and permitted to cool to normal temperature without further delay, because after cooling to 1000 degrees F. they would not again become hard without the application of more heat.

The above method is excellent for annealing high-speed steel on a large scale. If it is desired to anneal only a few small pieces of this grade of steel rapidly, they can be "water annealed," by a method similar to that used for carbon steels; the temperature to which the steel is raised, however, is not as high as for carbon steel. In water annealing, the piece to be annealed is gradually and uniformly heated to 760 degrees F. It is then taken from the furnace and plunged into a bath of pure water, previously heated to a temperature of 150 degrees F., where it is permitted to cool until reduced to the temperature of the bath. Afterwards the steel can be drilled, filed, or machined into any form with little difficulty. The more care devoted to the heating, the better the results will be. To heat rapidly will induce internal strains and greatly increase the risk of breakage when the pieces are plunged into the water bath.

Another annealing method which differs considerably from those outlined on the preceding pages is used by a well-known tool manufacturer. While doubt has been expressed as to its practicability, it is claimed to give good results. There is only one objection; the pieces annealed will scale off somewhat, but as the surface is generally machined anyway, this objection is — for many classes of work — of no importance. The method is as follows:

Pack the tools to be annealed directly in the oven, one on top of the other, the furnace being entirely filled if necessary. Then heat the furnace to a temperature not exceeding 1700 or 1750 degrees F. It should not require more than three hours for the furnace to reach this heat, which is then maintained for about two hours more, or until the temperature of all the tools has been raised to that of the furnace itself. (When smaller pieces are to be annealed, it is, however, sufficient to maintain the heat for about one hour.) Then shut off the heat and at the same time close all holes, such as burner and draft holes, as carefully as possible, and let the tools cool off in the furnace. This cooling takes place very much quicker than when the first-mentioned method is used, because the tools are not packed, and, hence, there is a saving in time not only in the heating but also in the cooling. The greater part of the expense of annealing is thus saved on account of the saving in fuel, and the elimination of the packing, packing materials and the boxes.

Chapter Five - Heat-Treatment of Alloy Steel

The rapid development of the automobile industry in America has awakened a quick, keen appreciation of the great importance of proper heat-treatment of steel. Scientific heat-treatment is quite as essential as the quality of the steel. Ordinary steel may acquire good physical qualities with proper heat-treatment, and the best of steel can be ruined by defective methods. There must be thoroughness in the various operations of annealing, hardening and tempering, for only treatment carried on with care makes uniformity of product possible.

The difference between ordinary and the best steel is great. For example, the elastic limit of ordinary steel is about 40,000 pounds per square inch, with a reduction in area of, say, 50 per cent. Nickel steel properly heat-treated has an elastic limit of 80,000 to 100,000 pounds per square inch of section, with a reduction in area of 50 per cent, or more. Brittleness does not follow proper heat-treatment, the enduring quality being increased in a greater ratio than the elastic limit. Consequently crystallization, fatigue, or whatever we name the cause of breakage, is less likely to develop in a properly heat-treated and tempered material than in an annealed and soft material. This fact, discovered in the laboratory and established in actual practice, is now commonly accepted by metallurgical experts, notwithstanding that it completely overturns previous general belief.

Another, until recently, commonly accepted belief was that strength and stiffness are coordinate, or "the stronger a piece of steel, the stiffer it is." To illustrate, it was thought that if one piece were twice as strong as another, it would bend only one-half as much under a given weight; but actual test has shown that a chrome-nickel steel having an elastic limit of 150,000 pounds or more per square inch of section, bends under a given load the same amount as a carbon steel specimen, and this condition holds true as long as the load is within the elastic limit of the weaker material. The elastic limit of a well heat-treated steel spring is about 150,000 pounds per square inch, but a spring can be made of soft steel. If it is not loaded beyond its elastic limit, the spring will return to its original shape after every deflection, but the deflection would not be sufficient to make a good spring. In fact, it would be hardly noticeable, and the spring, of course, would be of little value. Between these extremes lie the steels used by the spring makers in the past.

Not only has the automobile industry forced the spring makers to depart from their old materials and methods, but the change extends all along the line. Assume that a 0.20 per cent carbon steel has been used to advantage for a given design of crankshaft, neither bending nor breaking through long-continued use, and that the bearing surfaces are as small in area as can be used without heating or excessive wear. A crankshaft of properly treated chrome-nickel steel, having an elastic limit four or five times as high as the 0.20 per cent carbon steel, would be no stiffer, but would have greatly increased life and reliability. The steel makers must be prepared to meet these new conditions. Sound knowledge of steel has spread fast among intelligent manufacturers; from the knowledge obtained in the laboratories established, where all materials are physically and chemically tested, they have learned to discriminate in selection. With known characteristics, a heat-treatment scientifically conducted is sure to bring results that make high-grade steels comparable with ordinary steels in about the ratio the latter, in turn, bear to cast iron.

The cost of the materials used in automobile construction amounts to about sixty per cent of the total cost of production. In view of this fact, the kind of material best suited for the more vital parts is highly important. In the following the composition and treatment of some of the most commonly used alloy steels are reviewed. The carbon steels used in automobile construction are also included, in order to give a complete review of the subject. The information given is mainly from a report by the Iron and Steel Division of the Society of Automobile Engineers, January, 1912, in which very complete specifications for the composition, heat-treatment and properties of various kinds of steel were given. Tables relating to the composition, heat-treatment,; etc., of carbon, nickel, nickel-chromium and chromium-vanadium steels are also given.

The steels specified may be of open hearth, crucible or electric manufacture, and must be homogeneous, sound and free from physical defects, such as pipes, seams, heavy scale or scabs, and surface and internal defects visible to the naked eye. The figures given in the tables for physical characteristics or properties of the various steels refer to sections common in automobile construction, that is, to bars from 1 to 1½ inch round. The high elastic limits can be obtained only on small sections with very careful heat-treatment, while the low elastic limits can be expected on heavy sections with less refined or severe heat-treatment.

Carbon Steels. — The 0.10 per cent carbon steel is usually known to the trade as soft, basic open-hearth steel, and is commonly used for seamless tubing, pressed steel frames and brake-drums, sheet steel brake-bands, etc. It is soft and ductile and will stand a great deal of deformation without cracking. In its natural or annealed condition it should not be used where a great deal of strength is required. The quality of this material, however, is improved by cold drawing or rolling. An important fact to remember is that this steel when cold drawn or rolled is returned to the characteristics of the annealed material by heating. This remark also applies to all materials the elastic limit of which has been increased by cold working.

The 0.10 per cent carbon steel in its natural or annealed state cannot be easily machined. It will tear badly in turning, threading and broaching operations. Heat-treatment has little effect upon it, and does not increase its strength but only the toughness. The heat-treatment which will produce some stiffness is to quench it in oil or water at a temperature of 1500 degrees F. No drawing is required. This steel will caseharden but is not as suitable for this purpose as 0.20 per cent carbon steel. This latter steel is often known to the trade as machine steel and is intended primarily for casehardening. It forges well and machines well, but is not suitable for screw machine work. Its particular use is for forged, machined and casehardened parts, where strength is not of especial importance. It can also be drawn into tubes and rolled into cold rolled forms and is a good frame material. It can be interchanged with o.io per cent carbon steel for cold pressed shapes.

Heat-treatment of 0.20 per cent carbon steel does not increase its strength to any degree, but causes a refinement of the grain after forging and increases the toughness; all that is necessary is to quench it in oil at 1500 degrees F. The casehardening treatment specified in the accompanying tables, is the most important treatment for this quality of steel. The heat-treatment specified as "*A*" is for parts which do not need to carry a great deal of load or withstand shock, but simply must have a hard surface. Heat-treatment "*B*" is for the parts which must not only be hard on the surface but also must possess strength, as, for example, gears, cam rollers, steering-wheel pivot-pins, etc.

The 0.30 per cent carbon steel is primarily a structural steel. It forges well, machines well, and responds to heat-treatment in regard to strength as well as toughness. It is used for such forgings as axles, driving-shafts, steering pivots and other structural parts. This quality of steel is not intended for case-hardening, but, by careful treatment, it may be safely casehardened, although it is used for this purpose only as an emergency. In that case, it should be given a double heat-treatment, followed by a drawing operation.

The 0.40 per cent carbon steel is a structural steel of greater strength than that previously mentioned. Its uses are more limited and generally confined to such parts as demand a high degree of strength and a considerable degree of toughness. It is commonly used for crankshafts, driving shafts and propeller shafts. It has also been used for transmission gears, but is not quite hard enough for casehardening, and when casehardened, not tough enough to make safe transmission gears. When properly annealed it machines well, but is not suitable for screw machine work. The 0.50 per cent carbon steel differs but little from that just described, although owing to its higher carbon content, it is somewhat harder to machine and also somewhat stronger.

The 0.80 per cent carbon steel is ordinarily known to the trade as spring steel, and is generally used for springs of light sections. The 0.95 per cent carbon steel is also generally used for springs. When properly heat-treated extremely good results are possible. The quenching temperature, as specified in heat-treatment "*F*" in the accompanying tables, should, if anything, be lower than that specified. Because of the high carbon content, the steel is used considerably for heavier types of springs.

Nickel Steel. — Nickel steel is the most generally used of the alloy steels. The best quality contains 0.20 to 0.25 per cent carbon, 3.50 per cent nickel, 0.50 to 0.80 per cent manganese, and not over 0.04 per cent sulphur and phosphorus. With carbon and nickel as given above, the manganese content ought never to exceed the limits mentioned. A slightly lower carbon content is often used for casehardening purposes, and a higher carbon percentage is much used for crankshafts. Nickel steel is usually made in the basic open-hearth furnace. It is an excellent steel for casehardening, and is easier to machine than other alloy steels. With regard to the use of all alloy steels it should be borne in mind that such steels must be heat-treated and not used in the annealed or natural condition. In the latter condition they are but slightly superior to plain carbon steels. In the heat-treated condition, however, a marked improvement in physical characteristics is shown.

The 0.15 per cent carbon nickel steel, the analysis of which is given in the accompanying tables, is suitable for carbonizing purposes. Steel of this character properly carbonized and heat-treated will produce a part with an exceedingly tough and strong core and a hard exterior. This steel can be used for structural purposes, but is not especially suitable for this purpose. It is intended for casehardened gears and for such other casehardened parts as require both strength and hardness. The 0.20 per cent carbon nickel steel may be used interchangeably with that just described. It is intended primarily for casehardening purposes, but may, with suitable heat-treatment, also be used for structural parts. The 0.25 per cent carbon nickel steel may also be casehardened successfully and is satisfactory for gears — either of the transmission or the rear axle bevel type. The treatment for carbonizing must be slightly modified to meet the increase in carbon content. It can also be used for many structural parts if subjected to heat-treatment "*H*" or "*K*."

The 0.30 per cent carbon nickel steel is primarily used for structural parts where strength and toughness are required, for example, such parts as axles, crankshafts, driving-shafts and transmission shafts. Wide variations as to elastic limits are possible by varying the quenching mediums — oil, water or brine — and by variations in the drawing temperatures. This material may be casehardened, but is not suitable for that purpose. The 0.35 per cent carbon nickel steel is very similar to that just described.

The 0.40, 0.45 and 0.50 per cent carbon nickel steels are not widely used, but are available for certain purposes. A greater hardness is obtainable in these steels than in those of the lower carbon contents, but as increased brittleness accompanies the greater hardness, the treatment given must be modified to meet these conditions. For example, the final quenching must be at a lower temperature in order to produce the desired toughness and other properties. The strength of these steels depends upon the heat-treatment and may be controlled closely over a wide range.

Nickel-chromium Steels. — There are three types of nickel-chromium steels in common use, known as low, medium and high nickel-chromium steels according to the percentages of nickel and chromium. Nickel-chromium steels are also made either with a high carbon content, and used for oil-hardened gears and springs, or with a low carbon content, in which case the steel is used for axles, shafts, forged parts, and casehardened gears. The high-carbon steel carries about 0.5 per cent of carbon, while the low carbon alloy carries 0.25 per cent. The nickel content is from 1 to 3.5 per cent, while the chromium varies from 0.30 to 1.5 per cent. A special nickel-chrome-tungsten steel is sometimes used for springs. Nickel-chromium steels possess excellent static qualities, but present difficulties in heat-treatment, forging and machining.

Silico-manganese and silico-chromium steels with medium and low carbon contents are used to a considerable extent abroad for springs and gears. Their relatively low cost favors their use, but they do not stand up well when subjected to shocks, and are too sensitive to heat-treatment. When handled with great care they give good results where the temperatures for the heat-treatment can be accurately gaged. Chromium steels with high carbon content are used to a considerable extent for balls and ball races.

In general, it may be said that the heat-treatment and properties of these steels are much the same as those of the plain nickel steels, except that the effects of the heat-treatment are somewhat augmented by the presence of chromium. The low nickel-chromium steels with carbon contents up to 0.20 per cent are intended primarily for casehardening, while those with carbon contents from 0.25 to 0.40 per cent are intended primarily for structural purposes. Those with carbon contents from 0.45 to 0.50 per cent may be used for gears and other structural parts where a high degree of strength and hardness is demanded and where toughness is not of first importance.

The medium nickel-chromium steels are of the same composition as the low nickel-chromium steels except that they contain more nickel and chromium. Their general usage is practically the same as already mentioned for the low nickel-chromium steels.

The high nickel-chromium steels require different heat-treatments from the other two types mentioned on account of the amount of nickel and chromium that they contain. Annealing before machining will be found necessary for these steels. The higher percentages of nickel and chromium make machining in a natural condition difficult. The steels with low carbon contents are casehardened the same as in the case of low nickel-chromium steels, and those with higher carbon contents are used for structural parts. In general, these steels are used for parts of an important character, and where unusual strength is demanded. The 0.45 per cent high nickel-chromium steel, for example, is used for gears where extreme strength and hardness are necessary. The carbon content is sufficiently high to cause the material to become hard enough to make a good gear when quenched without being casehardened. This steel, however, is difficult to forge. During the forging operation it should be kept at a high or plastic heat and should not be hammered or worked after dropping to ordinary forging temperatures, as cracking is liable to follow. On the other hand, too high a temperature is not advisable, as the steel then becomes red-short and breaks.

Chrome-vanadium Steels. — The chrome-vanadium alloy steels are preferably made in the crucible or electric furnace, although the open-hearth process is also much used for the purpose. The open-hearth product, however, is somewhat uncertain, and while springs of steel made by this process may be better than those made from ordinary crucible steel, they cannot be compared with springs made of crucible chrome-vanadium steel. For excellent quality the latter product constitutes the highest attainment of the steel makers' art, and it seems that for springs no material is better suited than this steel.

Chrome-vanadium steel made with a high carbon content is suitable for oil-hardened gears and springs. When made with a low carbon content it is used for casehardened gears, and, when oil-quenched and annealed, for axles, shafts and steering knuckles. When a better material than the best nickel steel is needed, the various kinds of chrome-vanadium steel are to be recommended. They can be easily forged and can be machined more readily than chrome-nickel steels of corresponding carbon percentages.

Chrome-vanadium steels are used for many automobile parts. They are used interchangeably with carbon steels, nickel steels and nickel-chromium steels. Those qualities which contain from 0.15 to 0.20 per cent carbon are intended primarily for casehardened parts, while those of from 0.25 to 0.50 per cent carbon are used for structural parts. The 0.25 per cent carbon steel may be casehardened, but is not suitable for this purpose. The 0.40 per cent carbon steel is of a very good quality, to be selected where a high degree of strength is desired coupled with a moderate measure of toughness. It is a first-class material for high-duty shafts. The 0.45 per cent carbon steel may be used for gears and springs. When used for structural parts, if an exceedingly high strength is desirable, heat-treatment "*T*" should be used instead of treatment "*U*." The 0.50 per cent carbon chrome-vanadium steel is suitable for springs and gears. The final drawing temperature must vary with the section of material being handled; it must be taken into account, for example, whether light spiral springs or heavy flat springs are being heat-treated.

Casehardened vs. Oil-hardened Gears. — Both casehardened and oil-hardened gears are largely used in automobile construction. As previously mentioned, the chrome-vanadium, chrome-nickel and silico-manganese alloys are made with both high and low carbon contents. The former contains about 0.45 to 0.60 per cent carbon and enough other hardening elements so that by merely quenching the steel in oil from a bright red heat, surface hardening is produced sufficient for ordinary wearing purposes, while the hardness does not penetrate deeply into the gear, but leaves a tough and strong core. The low carbon alloy steels, with about 0.20 per cent carbon, require to be casehardened in order to produce a sufficiently hard surface for wearing purposes. The observations of many engineers have led them to prefer the casehardened gear, the following conclusions being based on the results of direct tests on thousands of gears.

1. The static strength of casehardened gears is equal to that of oil-hardened gears, assuming that in both cases steel of the same class of appropriate composition has been used, and the respective heat-treatments have been equally well and properly conducted.

2. Direct experiments prove that casehardened gears resist shocks better than oil-hardened gears.

3. The casehardened gear resists wear incomparably better, although it is perhaps not as silent in action.

The strong objection to the casehardening is in nine cases out of ten doubtless due to the fact that the casehardening operation is not properly understood. The depth of the hard case or covering, the time and temperature required to produce certain results, and the exact control of the conditions, together with an accurate knowledge of the material to be treated, are factors which enter into successful casehardening.

To obtain the best results in casehardening ordinary carbon steel, the following rules should be observed. Steel containing less than 0.12 per cent of carbon, and with a low percentage of manganese (less than 0.30 per cent) should be used; the casehardening should be accomplished by a material of a definitely known chemical composition, such as a mixture of 60 per cent charcoal and 40 per cent barium carbonate, and at a temperature between 1560 to 1920 degrees F. The higher the temperature, the more rapid will be the casehardening. After the casehardening operation, allow the steel to cool down to about 1 100 degrees F. Then reheat the work to be casehardened, and quench it at 1650 degrees F. This heating and quenching has the effect of toughening the center, but the outside will be coarse-grained and brittle; therefore heat the material a second time to 1470 degrees F. to render the outside non-brittle.

This procedure is more elaborate than that most commonly used, in which pieces are dumped directly from the casehardening boxes into water. The process, however, can be somewhat modified if one uses a good grade of nickel steel, low in carbon, and after having casehardened it at the appropriate temperature, permits the material to cool off in the boxes before reheating and quenching. In this case, if the material is reheated but once to 1470 degrees F. the result will be fully equal to or better than those obtained by the most careful annealing and double quenching of ordinary carbon steels. It is, however, better to give a double quenching, as then extraordinary toughness and wearing qualities are obtained.

An ideal way of making a nickel steel gear consists in first annealing the blank, then rough machining it approximately to size, and then re-annealing before taking the last finishing cut. The gears are then packed in a mixture, as mentioned, heated to a temperature of about 1625 to 1650 degrees F., carbonizing to a depth of about 1/64 to 1/32 inch. The gears are then permitted to cool in the boxes, are heated to 1500 degrees F., and quenched in a hot brine or calcium-chloride solution, and finally reheated to 1375 or 1400 degrees F.

Composition, Heat-treatment & Properties of Carbon and Alloy Steels - 1

(Information given is based on, and condensed from, the report of the Iron and Steel division of the Society of Automobile engineers, January, 1912.

	Kind of Steel and Nominal Carbon Content	Carbon, Per cent	Manganese, Per cent	Phosphorus,* Per cent	Sulphur,* Per cent	Nickel, Per cent
Carbon Steels	0.10 Carbon	0.05–0.15 (0.10)	0.30–0.60 (0.45)	0.04	0.04	....
	0.20 Carbon	0.15–0.25 (0.20)	0.50–0.80 (0.65)	0.04	0.04	....
	0.30 Carbon	0.25–0.35 (0.30)	0.50–0.80 (0.65)	0.04	0.04	....
	0.40 Carbon	0.35–0.45 (0.40)	0.50–0.80 (0.65)	0.04	0.04	....
	0.50 Carbon	0.45–0.55 (0.50)	0.50–0.80 (0.65)	0.04	0.04	....
	0.80 Carbon	0.75–0.90 (0.80)	0.25–0.50 (0.35)	0.04	0.04	....
	0.95 Carbon	0.90–1.05 (0.95)	0.25–0.50 (0.35)	0.04	0.04	....
Nickel Steels	0.15 Carbon Nickel	0.10–0.20 (0.15)	0.50–0.80 (0.65)	0.04	0.04	3.25–3.75 (3.50)
	0.20 Carbon Nickel	0.15–0.25 (0.20)	0.50–0.80 (0.65)	0.04	0.04	3.25–3.75 (3.50)
	0.25 Carbon Nickel	0.20–0.30 (0.25)	0.50–0.80 (0.65)	0.04	0.04	3.25–3.75 (3.50)
	0.30 Carbon Nickel	0.25–0.35 (0.30)	0.50–0.80 (0.65)	0.04	0.04	3.25–3.75 (3.50)
	0.35 Carbon Nickel	0.30–0.40 (0.35)	0.50–0.80 (0.65)	0.04	0.04	3.25–3.75 (3.50)
	0.40 Carbon Nickel	0.35–0.45 (0.40)	0.50–0.80 (0.65)	0.04	0.04	3.25–3.75 (3.50)
	0.45 Carbon Nickel	0.40–0.50 (0.45)	0.50–0.80 (0.65)	0.04	0.04	3.25–3.75 (3.50)
	0.50 Carbon Nickel	0.45–0.55 (0.50)	0.50–0.80 (0.65)	0.04	0.04	3.25–3.75 (3.50)

[*] Phosphorus and sulphur not to exceed values given. Values within parentheses are preferred percentages.

Carbon Content	Heat-treatment†	Elastic Limit, Pounds per Square Inch		Reduction in Area, Per cent		Elongation in 2 inches, Per cent	
		Annealed	Heat-treated	Annealed	Heat-treated	Annealed	Heat-treated
0.10	Quench at 1500° F.	28,000–36,000	....	55–65	....	30–40	....
0.20	*A* or *B*	30,000–40,000	40,000–75,000	45–60	15–60	25–35	15–35
0.30	*C* or *D*	35,000–45,000	40,000–80,000	40–55	30–60	20–30	10–30
0.40	*E*	40,000–50,000	45,000–100,000	40–50	25–55	20–25	5–25
0.50	*E*	45,000–60,000	50,000–110,000	30–40	15–50	15–20	5–20
0.80	*F*	....	90,000–160,000	....	....	....	....
0.95	*F*	....	90,000–160,000	....	....	....	....
0.15	*G* *H* or *K*	35,000–45,000	(*H* or *K*) 40,000–80,000	45–65	40–65	25–35	15–35
0.20	*G* *H* or *K*	40,000–50,000	(*H* or *K*) 50,000–125,000	40–65	40–65	20–30	10–25
0.25	*G* *H* or *K*	40,000–50,000	(*H* or *K*) 60,000–130,000	40–60	30–60	20–30	10–25
0.30	*H* or *K*	45,000–55,000	65,000–150,000	35–55	25–55	15–25	10–25
0.35	*H* or *K*	45,000–55,000	65,000–160,000	35–55	25–55	15–25	10–25
0.40	*H* or *K*	55,000–70,000	70,000–200,000	30–50	15–55	15–25	5–20
0.45	*H* or *K*	55,000–70,000	70,000–200,000	30–50	15–55	15–25	5–20
0.50	*H* or *K*	55,000–70,000	70,000–200,000	30–50	15–55	15–25	5–20

[†] See Table 4 for detailed description of heat-treatment.

Composition, Heat-treatment & Properties of Carbon and Alloy Steels - 2

	Kind of Steel and Nominal Carbon Content	Carbon, Per cent	Manganese, Per cent	Phosphorus,* Per cent	Sulphur,* Per cent	Nickel, Per cent
Low Nickel-chromium Steels	0.15 Carbon Low Ni-Cr	0.10–0.20 (0.15)	0.50–0.80 (0.65)	0.04	0.04	1.00–1.50
	0.20 Carbon Low Ni-Cr	0.15–0.25 (0.20)	0.50–0.80 (0.65)	0.04	0.04	1.00–1.50
	0.25 Carbon Low Ni-Cr	0.20–0.30 (0.25)	0.50–0.80 (0.65)	0.04	0.04	1.00–1.50
	0.30 Carbon Low Ni-Cr	0.25–0.35 (0.30)	0.50–0.80 (0.65)	0.04	0.04	1.00–1.50
	0.35 Carbon Low Ni-Cr	0.30–0.40 (0.35)	0.50–0.80 (0.65)	0.04	0.04	1.00–1.50
	0.40 Carbon Low Ni-Cr	0.35–0.45 (0.40)	0.50–0.80 (0.65)	0.04	0.04	1.00–1.50
	0.45 Carbon Low Ni-Cr	0.40–0.50 (0.45)	0.50–0.80 (0.65)	0.04	0.04	1.00–1.50
	0.50 Carbon Low Ni-Cr	0.45–0.55 (0.50)	0.50–0.80 (0.65)	0.04	0.04	1.00–1.50
Medium Nickel-chromium Steels	0.15 Carbon Med. Ni-Cr	0.10–0.20 (0.15)	0.30–0.60 (0.45)	0.04	0.04	1.50–2.00 (1.75)
	0.20 Carbon Med. Ni-Cr	0.15–0.25 (0.20)	0.30–0.60 (0.45)	0.04	0.04	1.50–2.00 (1.75)
	0.25 Carbon Med. Ni-Cr	0.20–0.30 (0.25)	0.30–0.60 (0.45)	0.04	0.04	1.50–2.00 (1.75)
	0.30 Carbon Med. Ni-Cr	0.25–0.35 (0.30)	0.30–0.60 (0.45)	0.04	0.04	1.50–2.00 (1.75)
	0.35 Carbon Med Ni-Cr	0.30–0.40 (0.35)	0.30–0.60 (0.45)	0.04	0.04	1.50–2.00 (1.75)
	0.40 Carbon Med. Ni-Cr	0.35–0.45 (0.40)	0.30–0.60 (0.45)	0.04	0.04	1.50–2.00 (1.75)
	0.45 Carbon Med. Ni-Cr	0.40–0.50 (0.45)	0.30–0.60 (0.45)	0.04	0.04	1.50–2.00 (1.75)

[*] Phosphorus and sulphur not to exceed values given. Values within parentheses are preferred percentages.

Carbon Content	Chromium, Per cent	Heat-treatment †	Elastic Limit, Pounds per Square Inch		Reduction in Area, Per cent		Elongation in 2 Inches, Per cent	
			Annealed	Heat-treated	Annealed	Heat-treated	Annealed	Heat-treated
0.15	0.30–0.75	G H or K	30,000–40,000	(H or K) 40,000–100,000	40–55	40–65	25–35	15–25
0.20	0.30–0.75	G H or K	30,000–40,000	(H or K) 40,000–100,000	40–55	40–65	25–35	15–25
0.25	0.30–0.75	H or K	40,000–55,000	50,000–125,000	35–50	25–55	20–30	10–25
0.30	0.30–0.75	H or K	40,000–55,000	50,000–125,000	35–50	25–55	20–30	10–25
0.35	0.30–0.75	H or K	45,000–60,000	55,000–150,000	30–45	25–50	15–25	5–20
0.40	0.30–0.75	H or K	45,000–60,000	55,000–150,000	30–45	25–50	15–25	5–20
0.45	0.30–0.75	K	55,000–70,000	60,000–175,000	30–50	20–45	15–25	5–15
0.50	0.30–0.75	K	55,000–70,000	60,000–175,000	30–50	20–45	15–25	5–15
0.15	0.75–1.25 (0.75)	G H or K	35,000–45,000	(H or K) 45,000–110,000	45–55	35–65	20–30	10–25
0.20	0.75–1.25 (1.00)	G H or K	35,000–45,000	(H or K) 45,000–110,000	45–55	35–65	20–30	10–25
0.25	0.75–1.25 (1.00)	H or K	40,000–50,000	55,000–130,000	40–55	30–60	20–30	10–25
0.30	0.75–1.25 (1.00)	H or K	45,000–50,000	55,000–150,000	35–50	30–55	15–25	10–25
0.35	0.75–1.25 (1.00)	H or K	45,000–55,000	60,000–160,000	35–50	30–55	15–25	5–20
0.40	0.75–1.25 (1.00)	H or K	50,000–60,000	60,000–175,000	35–45	25–50	15–25	5–20
0.45	0.75–1.25 (1.00)	K	55,000–65,000	100,000–200,000	35–45	20–35	15–25	0–15

[†] See Table 4 for detailed description of heat-treatment.

Composition, Heat-treatment & Properties of Carbon and Alloy Steels - 3

	Kind of Steel and Nominal Carbon Content	Carbon, Per cent	Manganese, Per cent	Phosphorus,* Per cent	Sulphur,* Per cent	Nickel, Per cent	Chromium, Per cent
High Nickel-chromium Steels	0.15 Carbon High Ni-Cr	0.10-0.20 (0.15)	0.30-0.60 (0.45)	0.04	0.04	3.25-3.75 (3.50)	1.25-1.75 (1.50)
High Nickel-chromium Steels	0.20 Carbon High Ni-Cr	0.15-0.25 (0.20)	0.30-0.60 (0.45)	0.04	0.04	3.25-3.75 (3.50)	1.25-1.75 (1.50)
High Nickel-chromium Steels	0.25 Carbon High Ni-Cr	0.20-0.30 (0.25)	0.30-0.60 (0.45)	0.04	0.04	3.25-3.75 (3.50)	1.25-1.75 (1.50)
High Nickel-chromium Steels	0.30 Carbon High Ni-Cr	0.25-0.35 (0.30)	0.30-0.60 (0.45)	0.04	0.04	3.25-3.75 (3.50)	1.25-1.75 (1.50)
High Nickel-chromium Steels	0.35 Carbon High Ni-Cr	0.30-0.40 (0.35)	0.30-0.60 (0.45)	0.04	0.04	3.25-3.75 (3.50)	1.25-1.75 (1.50)
High Nickel-chromium Steels	0.40 Carbon High Ni-Cr	0.35-0.45 (0.40)	0.30-0.60 (0.45)	0.04	0.04	3.25-3.75 (3.50)	1.25-1.75 (1.50)
High Nickel-chromium Steels	0.45 Carbon High Ni-Cr	0.40-0.50 (0.45)	0.30-0.60 (0.45)	0.04	0.04	3.25-3.75 (3.50)	1.25-1.75 (1.50)
Chrome-vanadium Steels	0.15 Carbon Cr-Va	0.10-0.20 (0.15)	0.50-0.80 (0.65)	0.04	0.04	....	0.70-1.10 (0.90)
Chrome-vanadium Steels	0.20 Carbon Cr-Va	0.15-0.25 (0.20)	0.50-0.80 (0.65)	0.04	0.04	....	0.70-1.10 (0.90)
Chrome-vanadium Steels	0.25 Carbon Cr-Va	0.20-0.30 (0.25)	0.50-0.80 (0.65)	0.04	0.04	....	0.70-1.10 (0.90)
Chrome-vanadium Steels	0.30 Carbon Cr-Va	0.25-0.35 (0.30)	0.50-0.80 (0.65)	0.04	0.04	....	0.70-1.10 (0.90)
Chrome-vanadium Steels	0.35 Carbon Cr-Va	0.30-0.40 (0.35)	0.50-0.80 (0.65)	0.04	0.04	....	0.70-1.10 (0.90)
Chrome-vanadium Steels	0.40 Carbon Cr-Va	0.35-0.45 (0.40)	0.50-0.80 (0.65)	0.04	0.04	....	0.70-1.10 (0.90)
Chrome-vanadium Steels	0.45 Carbon Cr-Va	0.40-0.50 (0.45)	0.50-0.80 (0.65)	0.04	0.04	....	0.70-1.10 (0.90)
Chrome-vanadium Steels	0.50 Carbon Cr-Va	0.45-0.55 (0.50)	0.50-0.80 (0.65)	0.04	0.04	....	0.70-1.10 (0.90)

[*] Phosphorus and sulphur not to exceed values given. Values within parentheses are preferred percentages.

Carbon Content	Vanadium,† Per cent	Heat-treatment‡	Elastic Limit, Pounds per Square Inch		Reduction in Area, Per cent		Elongation in 2 inches, Per cent	
			Annealed	Heat-treated	Annealed	Heat-treated	Annealed	Heat-treated
0.15	....	*L*	40,000–50,000	....	45–60	....	20–25	....
0.20	....	*L* *M* or *P*	40,000–50,000	(*M* or *P*) 50,000–125,000	45–60	30–65	20–25	5–20
0.25	....	*M* or *P*	40,000–50,000	60,000–140,000	45–60	30–65	20–25	5–20
0.30	....	*M* or *P*	45,000–55,000	60,000–175,000	40–55	30–60	15–25	5–20
0.35	....	*M* or *P*	45,000–55,000	60,000–175,000	40–55	30–60	15–25	5–20
0.40	....	*P*	50,000–60,000	65,000–200,000	40–50	20–50	15–25	2–15
0.45	....	*Q*	50,000–60,000	150,000–250,000	40–50	15–25	15–25	2–15
0.15	0.12 (0.18)	*S* or *T*	35,000–45,000	(*T*) 50,000–90,000	50–70	40–70	25–30	10–25
0.20	0.12 (0.18)	*T*	40,000–50,000	55,000–100,000	50–65	45–65	20–30	10–25
0.25	0.12 (0.18)	*T*	40,000–50,000	55,000–100,000	50–65	45–65	20–30	10–25
0.30	0.12 (0.18)	*T*	45,000–55,000	60,000–150,000	50–60	25–55	20–25	5–15
0.35	0.12 (0.18)	*T*	45,000–55,000	60,000–150,000	50–60	25–55	20–25	5–15
0.40	0.12 (0.18)	*T*	50,000–60,000	65,000–175,000	45–55	15–50	15–25	2–15
0.45	0.12 (0.18)	*U*	55,000–65,000	150,000–200,000	40–55	10–25	15–25	2–10
0.50	0.12 (0.18)	*U*	60,000–70,000	150,000–225,000	35–50	15–35	15–20	2–10

[†] Vanadium content not to be less than 0.12 per cent.
[‡] See Table 4 for detailed description of heat-treatment.

Heat-treatment of Carbon and Alloy Steels – 4

Heat-treatments specified for various steels listed in preceding Tables 1, 2 and 3.

Heat-treatment A

After forging or machining:
1. Carbonize at a temperature between 1600° F. and 1750° F. (1650°–1700° F. desired).
2. Cool slowly or quench.
3. Reheat to 1450°–1500° F. and quench.

Heat-treatment B

After forging or machining:
1. Carbonize at a temperature between 1600° F. and 1750° F. (1650°–1700° F. desired).
2. Cool slowly in the carbonizing mixture.
3. Reheat to 1500°–1550° F.
4. Quench.
5. Reheat to 1400°–1450° F.
6. Quench.
7. Draw in hot oil at a temperature which may vary from 300° to 450° F., depending upon the degree of hardness desired.

Heat-treatment C

After forging or machining:
1. Heat to 1475°–1525° F.
2. Quench.
3. Reheat to 600°–1200° F. and cool slowly.

Heat-treatment D

After forging or machining:
1. Heat to 1500°–1550° F.
2. Quench.
3. Reheat to 1400°–1450° F.
4. Quench.
5. Reheat to 600°–1200° F. and cool slowly.

Heat-treatment E

After forging or machining:
1. Heat to 1500°–1550° F.
2. Cool slowly.
3. Reheat to 1400°–1450° F.
4. Quench.
5. Reheat to 600°–1200° F. and cool slowly.

Heat-treatment F

After shaping or coiling:
1. Heat to 1425°–1475° F.
2. Quench in oil.
3. Reheat to 400°–800° F. in accordance with degree of temper desired, and cool slowly.

Heat-treatment G

After forging or machining:
1. Carbonize at a temperature between 1600° F. and 1750° F. (1650°–1700° F. desired).
2. Cool slowly in carbonizing material.
3. Reheat to 1450°–1525° F.
4. Quench.
5. Reheat to 1300°–1400° F.
6. Quench.
7. Reheat to a temperature from 250°–500° F. (in accordance with the necessities of the case) and cool slowly.

Heat-treatment H

After forging or machining:
1. Heat to 1500°–1550° F.
2. Quench.
3. Reheat to 600°–1200° F. and cool slowly.

Heat-treatment K

After forging or machining:
1. Heat to 1500°–1550° F.
2. Quench.
3. Reheat to 1300°–1400° F.
4. Quench.
5. Reheat to 600°–1200° F. and cool slowly.

Heat-treatment L

After forging or machining:
1. Carbonize at a temperature between 1600° F. and 1750° F. (1650°–1700° F. desired).
2. Cool slowly in the carbonizing mixture.
3. Reheat to 1400°–1500° F.
4. Quench.
5. Reheat to 1300°–1400° F.
6. Quench.
7. Reheat to 250°–500° F. and cool slowly.

Heat-treatment of Carbon and Alloy Steels – 4 (*Continued*)

Heat-treatments specified for various steels listed in preceding Tables 1, 2 and 3.

Heat-treatment M

After forging or machining:

1. Heat to 1450°–1500° F.
2. Quench.
3. Reheat to a temperature between 500° F. and 1250° F. and cool slowly.

Heat-treatment P

After forging or machining:

1. Heat to 1450°–1500° F.
2. Quench.
3. Reheat to 1375°–1425° F.
4. Quench.
5. Reheat to a temperature between 500° F. and 1250° F. and cool slowly.

Heat-treatment Q

After forging:

1. Heat to 1475°–1525° F. (Hold at this temperature one-half hour, to insure thorough heating.)
2. Cool slowly.
3. Reheat to 1450°–1500° F.
4. Quench.
5. Reheat to 250°–550° F. and cool slowly.

Heat-treatment S

After forging or machining:

1. Carbonize at a temperature between 1600° F. and 1750° F. (1650°–1700° F. desired).
2. Cool slowly in the carbonizing mixture.
3. Reheat to 1600°–1700° F.
4. Quench.
5. Reheat to 1475°–1550° F.
6. Quench.
7. Reheat to 250°–550° F. and cool slowly.

Heat-treatment T

After forging or machining:

1. Heat to 1600°–1700° F.
2. Quench.
3. Reheat to some temperature between 500° F. and 1300° F. and cool slowly.

Heat-treatment U

After forging:

1. Heat to 1525°–1600° F. (Hold for about one-half hour.)
2. Cool slowly.
3. Reheat to 1650°–1700° F.
4. Quench.
5. Reheat to 350°–550° F. and cool slowly.

-----

and quenched in oil. The temper need not be drawn.

It has been stated that drop forging is detrimental to the strength of steel. This is true only when the drop forging is done under improper conditions, by inexperienced men, and with inadequate facilities. On the other hand, drop forgings made by reputable concerns are claimed to possess a tensile strength of from 10 to 12 per cent greater than that of the bar from which they are made; for example, mild steels, having a tensile strength of from 55,000 to 58,000 pounds per square inch, may have this strength increased to 62,000 pounds by drop forging.

Necessity of Heat-treatment of Alloy Steels. — While the best alloy steels are none too good for most of the parts in automobile construction, their qualities will not become pronounced unless they receive proper heat-treatment. It is waste of money to buy good alloy steels without knowing how to properly treat them to bring forth their exceptional qualities. For gaging the heat a pyrometer is necessary, but it is too often supposed to take care of itself. The best pyrometer of the thermo-couple type should be regularly inspected.

The heat-treatment operations depend upon established scientific facts, and a lack of appreciation of this causes many people to buy high-priced alloy steels from which they get no better results than from carbon steel properly handled. As an example of the effect of heat-treatment may be mentioned a chrome steel which in its rolled condition had an elastic limit of 158,000 pounds, 5 per cent elongation, and 9.4 per cent reduction in area. The same steel, oil tempered and annealed, had an elastic limit of 153,000 pounds, 14 per cent elongation and 52 per cent reduction in area. In other words, the material was transformed from brittle to tough without appreciably affecting its elastic limit.

Chapter Six - Heat-Treatment of Steel by the Electric Furnace

In properly managed shops, the heat-treatment of steel is today receiving thorough attention. To produce a tool of such high quality that it will give several times the service of a tool that has not been properly heat-treated, is an important factor in shop economy. To accomplish this result it is necessary to know the laws governing the hardening of steel. If we clearly understand the causes underlying the changes which take place when steel is subjected to various heat-treatments, we have the basis for a positive control of the quality of the finished product. "Electric heat" is a new and important means to this end.

We heat and quench a tool because we want it to be harder. In every case the object of this treatment is to change, in some degree, certain of the physical properties of the steel. The effect of heat upon a piece of steel depends on the nature of the steel — that is, upon its composition, its form or external shape, and its internal structure. The raising of the temperature of the piece sufficiently will produce a change in the form, so as to increase its volume by causing a lengthening in some directions. In the structure of the steel itself this may introduce mechanical strains in the fibers. Mostly, however, these are but temporary changes, and, with proper heat-treatment, they disappear when the piece is again cooled.

Changes in the chemical arrangement of the elements composing the steel are produced when the temperature of the piece is raised to a sufficient degree. These are the changes that are effective in hardening and tempering. If the piece, once heated to a sufficient temperature to produce hardening, is allowed to cool very slowly, these "changes" of chemical arrangements revert to their original condition; but if the piece is cooled quickly — quenched — immediately upon removing it from the source of heat, the changes are made permanent.

For steels of different composition, that is, made up of either different elements or of different proportions of the same elements (iron, carbon, etc.), there are, as explained in Chapter One, different critical temperatures at which these changes take place. Corresponding differences in the heat-treatment are, therefore, necessary to produce the best results. Even two pieces of the same steel which vary greatly in their form must be treated differently. This is true also of two pieces of steel the composition and form of which may be identical, but the ultimate use of which may be different.

Examples of all three of these conditions occur constantly in shop practice. The intelligent hardener knows that a complicated die must be handled differently from a straight lathe tool, and a shaper tool for soft metal need not be as hard as one for hard metal, even though all of these pieces be made from the same steel. The next step ahead is to treat different kinds of steels according to their particular requirements. Various high-speed steels, high-carbon steels, and low-carbon steels are given individual treatment. Each is subjected to the conditions of heating and subsequent handling that will bring out the maximum of its useful characteristics. To meet the wide range of requirements and to avoid the losses of tools spoiled in hardening, there are three factors in the heating of steel which should be observed: First, the quality of the heat; second, its uniformity; and third, its degree.

Under ideal conditions the steel would be subjected to a heat effect only, as this alone is necessary to produce the desired changes. In practice, however, it is difficult to produce heat of the necessary intensity without having the quality of it impaired by the presence of flames or oxidizing gases and sulphur and other injurious fumes. Freeing the heat from such attendant defects would obviously greatly improve the quality of the finished product; also the more uniformly heat can be applied to all parts of the piece, the more uniform will be the hardening or tempering. These two factors are positive and constant in their desirability for all ordinary work. The third factor — the degree of heat — is the variable quantity. It is this factor, the temperature, that is chiefly manipulated to meet the requirements of different steels and of the same steels for different purposes. While other conditions have a certain influence, the temperature is the controlling factor in all heat-treatment. The evil effects of too high a temperature — the common failing — are well understood.

The length of time during which the steel is actually heated is an important point closely connected to that of the temperature. With a heating chamber of sufficient size to supply the necessary heat to the piece, the internal change in the steel that results in hardening can be effected, in general, in one of two ways; either the piece may be heated for a short time at a relatively high temperature, or for a longer interval at a lower temperature. Both may produce hardness. Fractures of small pieces of steel which after preheating were heated for but thirty seconds at a temperature 300 or 400 degrees higher than the hardening point, often compare very favorably with similar fractures of the same steel heated for four or five minutes at a temperature but little above the critical point. Thus the quantity of heat absorbed by the piece being treated is seen to be practically a product of the temperature of the heating chamber and the time the piece is left in it. In other words, the time of heating varies inversely as the temperature of the chamber in which the piece is heated. This still further emphasizes that "temperature should be the controlling factor," because, of the two extremes, the ordinary dangers of burning the steel on account of too high a temperature, and of causing it to crack due to too rapid, irregular heating, are far greater than those of "over-soaking" at lower temperature. To heat at the lower temperature is plainly the safe course, due both to its cutting down the grave danger of over-heating and to the greater uniformity with which heat is absorbed by the piece.

It is clear, then, that the heat-treatment of a particular steel can be greatly improved by definitely knowing beforehand the correct temperature at which it should be hardened. Also, when a large number of tools of approximately uniform sizes and shapes are being handled, the time necessary for proper heat absorption should first be determined, using an experimental tool that represents a fair sample. It will be found that the time element varies practically in proportion to the thickness of the steel. From the furnace standpoint, therefore, the accurate and flexible control of the temperature is a most important consideration. A positive means, such as a pyrometer, for indicating at any time just what the temperature is, becomes, of course, an incidental requirement.

Until recently, the only known way of producing heat of the required intensity was by combustion — the burning of some fuel. The attendant disadvantages of this are well known. The crude open coal forge is capable of heating the steel, but leaves much to be desired as regards the quality of the heat, its uniformity, and the temperature control. In order to produce heat at all, the carbon in the coal must be combined with the oxygen of the air, and a strongly oxidizing flame is unavoidable. The steel exposed to this action, or to the inevitable results of it, suffers accordingly. The coke-burning furnace offered some improvements, but only in detail. Now there are highly-perfected furnaces for burning oil and gas, and some of these offer still further advances, but the principle at the basis of all of these is the same — there must be a "burning" process to produce the heat; oxidation must be present with all fuel-combustion furnaces.

Through what means, then, may we obtain the proper quality of heat, uniformly applied, and of the right degree? The electric furnace for the heating of steel brings the answer. It overcomes most of the objections to the "combustion process" by introducing a new principle.

Electric Heat. — The heat of the electric furnace is produced in an entirely different way from that of the process of combustion. Electric heat can be produced by means of the electric arc, as in the arc lamp, and by the resist-

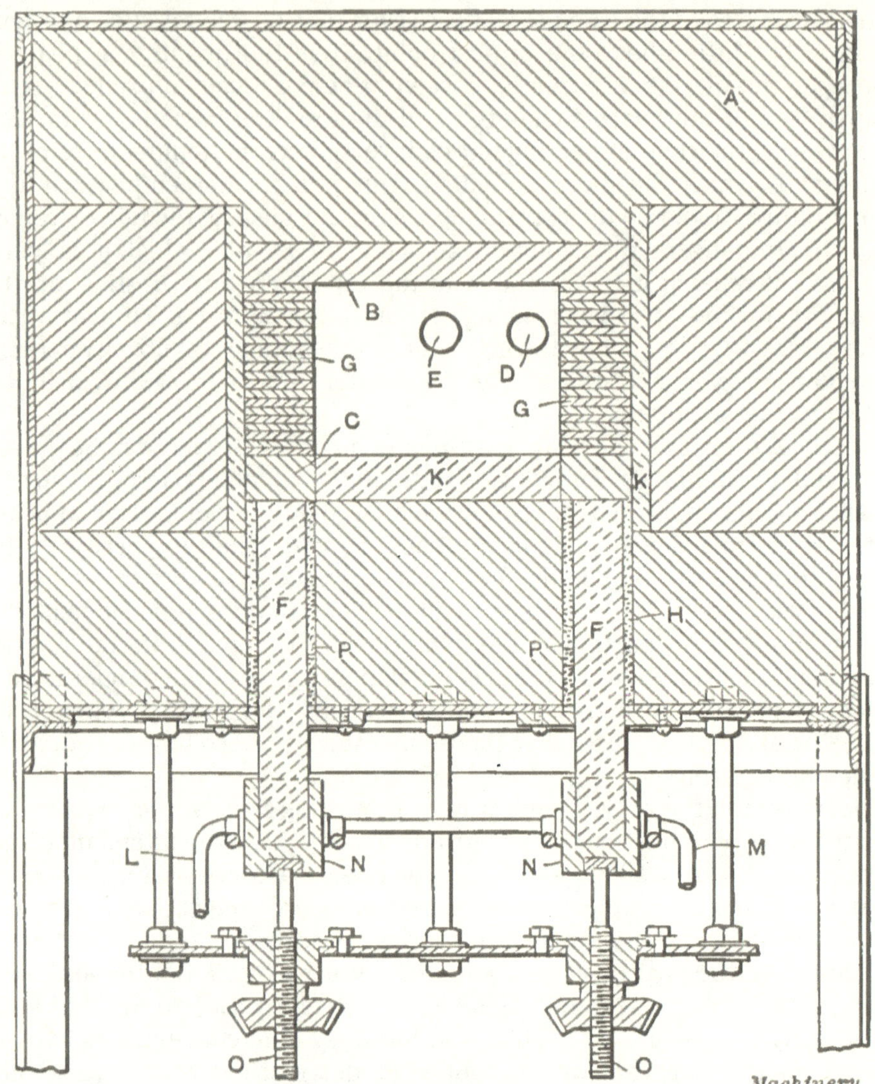

Fig. 1. Section through Electric Furnace

ance of a conductor, as in the incandescent lamp. It is the latter principle — due to its greater flexibility and convenience — that was utilized by Albert L. Marsh in the electric furnace developed by the Hoskins Mfg. Co., Detroit, Mich., for the heat-treatment of steel. Fig. 1 shows a section of a furnace of one of the larger sizes, the chamber in this being i8 inches deep (front to back), 12 inches wide and 8 inches high. The relation of the various constructional parts is clearly shown; *A* shows the fireclay insulation; *B*, the carbon connector plates; *C*, the graphite bottom plates; *D*, the draft hole; *E*, the pyrometer hole; *F*, the electrodes; *G*, the resistor plates; *H*, fire sand; *K*, cement filling; *L*, the inlet for the water used for cooling the electrode clamps; *M*, the outlet for this water; *N*, the electrode clamps; and *O*, the pressure regulating screws. The electrodes are surrounded by asbestos at *P*.

The full length of the side walls and the entire roof of the chamber are formed by the heating elements; the walls are composed of a series of thin carbon plates resting on the top of a heavy block of the same material, and the roof, of a thick graphite plate connecting these two columns at the top. One graphite electrode projects up to the middle of each side-wall plate and connects electrically, through water-cooled clamps at the lower end, with the source of energy. The chamber floor is of cement. Outside of the carbon plates there is a lining of the same material. This lining, with a carefully designed backing of heat-resisting material, retains the heat developed within the furnace. The counterweighted door fitted with a peep-hole serves as a quick access to the chamber, while in the rear wall are holes for the insertion of a pyrometer tube and for draft regulation. A rigid enclosing case of steel holds all parts securely.

The principle of operation is simple. A heavy low-voltage electric current is supplied through the electrodes to the resistor plates forming the side walls of the working chamber. Heat is generated here, due to the resistance offered by these plates to the passage of the current. The electrical "resistivity" of the carbon causes each plate to heat exactly as the carbon filament in the incandescent lamp "lights" when the current is turned on. In addition to this action, advantage is taken, in the furnace, of a second form of electrical resistance — that of the contact of one plate with another. This may be readily varied by altering the mechanical pressure on the plate columns by means of the hand-screws. The turning of these, changes the resistance of the circuit and hence the resulting temperature produced.

While the furnace is "electric" in its nature, it is not at all necessary that the hardening man handling it be an electrician. The simple electrical features of the furnace are quickly grasped. It is also safe, both because it practically eliminates the fire hazard and because it brings a corresponding protection to the operator. Normal working temperatures are acquired in a little over an hour's time after the switch has been closed. An average of 12½-kilowatt energy consumption will maintain the chamber at approximately 2250 degrees F.; higher temperatures, up to 2500 degrees F., which is even above the requirements of high-speed steels, or lower, as desired, may be obtained by increasing or decreasing the energy supply.

The life of the various parts of the heating unit is shown to be from 150 to 200 operating hours for the sidewall resistor plates, and 500 hours for the electrodes. Based on ten-hour day operation, it is found that the upkeep cost of these items, even on this larger-sized furnace, is less than 40 cents a running day. The rest of the furnace does not depreciate rapidly.

The Advantages of Electric Heating. — The atmosphere in the heating chamber of the electric furnace is inherently "reducing" in its nature, due to the fact that the hot carbon plates absorb all of the atmospheric oxygen. By raising the door slightly, and opening the draft-hole at the rear, a slight current of air may be admitted which will counteract this tendency. Leaving the door open slightly more would allow an excess of air to enter, so that an oxidizing atmosphere could be produced. Between the extreme points fine shades of atmospheric conditions can be obtained. Thus the quality of the heat can be absolutely and easily regulated.

Because of the arrangement possible with the electrical resistor, the heat may be generated within the working chamber itself. In the furnace described, the very walls of this chamber constitute the heat-generating device. The "resistivity" of the carbon plates is uniform — the same electric current runs through them all — with the result that an equal radiation of heat into the chamber takes place from practically every point in the walls.

In any type of furnace the temperature is varied, within limits, by varying the amount of energy transformed into heat. The regulation of the energy supply thus becomes the means of the temperature control. The electric energy control lends itself with exceeding exactness to meeting this principle. In the furnace described both a very fine and a wide regulation of temperature may be obtained by slight variations in the mechanical pressure between the carbon plates.

Commercial Importance. — Mr. Samuel S. Roberts, testing engineer of the Carnegie Steel Co., has been carefully investigating the subject of steel heating furnaces. In an interesting report of a series of tests which he, together with a number of steel experts, recently made on the heat-treatment of carbon and high-speed steel tools in the electric furnace, he says in part:

"A realization of the inadequacy of the prevailing furnace designs usually employed for the specific purpose of hardening and tempering specially formed tools of high-speed steel, such as formed milling and gear cutters, twist drills, taps, threading dies, reamers and other tools that do not permit of being ground to shape after being hardened, and where any melting or fusing of cutting edges must be prevented, has prompted the large tool steel consumers to welcome the advent of refined heating appliances, whereby the destructive influences hitherto encountered are eliminated. The modern electric resistance furnace, with its perfect heat control, evenly distributed heat maintenance at any desired point, reducing atmosphere, absence of all products of combustion, and thermo-electric pyrometer for measuring the temperature, offers not only the most attractive method whereby the consumers of tool steel are insured maximum efficiency, but has caused the science of treating the rapid cutting tools to take a long step forward."

Practical Application. — As to the cost of operation, it is a demonstrated fact that the higher the temperature it is desired to produce, the lower the cost of electric heat in comparison with fuel heat. At the lower ranges, considering only the production of the necessary heat alone and with electric power at the usual commercial rates, heat from this source costs considerably more than heat from fuels, especially the cheaper ones. Where water-power supplies of electric current are available, this ratio decreases in favor of electric heat; but the cost of producing the energy for the heat-treatment is only a part of that of the whole operation involved. When we consider that this broad factor includes the resulting service of the finished tool, as well as the labor, material and overhead charges to produce it, we see how comparatively small this part is. It is from such a view of the entire cost of production that the improved hardening of steel in electric heat is seen to be a real economy. The electric furnace, due to its advantages, makes possible a higher quality of product than is possible with fuel heating.

The use of electric furnaces for hardening and tempering steel has passed through the experimental stage and there are several types now on the market that have been made commercial successes. About 1910, an electric furnace was placed on the market that uses a molten salt bath in which to heat the steel to any of the hardening temperatures used for carbon, alloy or high-speed steels. This electric salt-bath furnace is also used for any drawing temperature above the flash point of oil, while an oil bath type is used to heat steel to drawing temperatures below this flash point. In the salt-bath furnaces, the electric current is sent directly through the salt by placing electrodes inside of the bath on opposite sides, as explained in detail in a following chapter, while in the oil-bath furnaces the oil tank is heated from the outside.

Since the introduction of the salt-bath furnace, the oven type of electric furnace has been perfected and installed by several manufacturers for commercial work. Small furnaces of this type have long been in use for laboratory and experimental work, but the principle used for heating, namely, sending the current through coiled resistance wires inside of the heating chamber, is too expensive for the larger furnaces used for commercial work, and the high temperatures required for high-speed steel cannot be reached without burning out the resistance wires. They are very successful, however, for small furnaces that are used at temperatures below 1800 degrees F.

One of the largest installations of the modern oven-type of electric furnaces is the battery of electric hardening furnaces that are in daily use at the Timken-Detroit Axle Co.'s plant. These were manufactured and installed by the Hoskins Mfg. Co. of Detroit, Mich. Each separate unit consists of a furnace, switchboard and transformer, which stands back of the switchboard. On the board is placed the line switch for turning on and off the current; a circuit breaker, adjusted to open the circuit automatically in case of an excessive flow of current; an ammeter to show the amount of current being used; and a pyrometer to indicate the temperature inside of the furnace. No rheostat is required, as the furnace can be adjusted to regulate the amount of current that is necessary to produce any given degree of temperature.

Heating Gears in the Electric Furnace. — The bulk of the work heated in these furnaces consists of the gears that enter into the construction of rear axle drives used on automobiles. It was found difficult to harden these large ring gears uniformly at accurate predetermined temperatures, without warpage, before the electric furnaces were installed, but now most of the difficulties have been overcome and good results are obtained. The gears are first pre-heated to about 1200 degrees F. in an ordinary kind of furnace. Four gears are then placed on the thick iron disk in the electric furnace, and heated to the correct temperature for quenching to harden. The disk is slowly revolved automatically, by mechanical means, to insure obtaining a uniform temperature around the entire circumference of the gears. To heat four carbon-steel gears, 11¼ inches in diameter, from the 1200 degrees of the preheating furnace to the 1500 degrees required to harden the gears, takes about twenty minutes. Prolonged heating does not harm the gears as the furnace is maintained at the correct hardening temperature and they will not be overheated, while the atmosphere is reducing and thus prevents the formation of any scale. If not left in long enough, however, they will not attain the temperature of the furnace and therefore the quenching will not give them the desired hardness.

Action of Carbon in Heated Steel. — It is a well-known fact that steel absorbs various elements with which it comes in contact, when the conditions are favorable. As is well known, this is the principle of the carbonizing process. Many impurities which injure some of the good qualities possessed by the metal are absorbed under certain conditions, and several elements that improve some qualities are absorbed under other conditions. Some of these conditions are known and can be controlled, while many others are unknown or only partially known.

Some beneficial elements, of which chromium is an example, can be injected into steel. Carbon flows through steel in a somewhat similar manner to the flow of electricity, although very much slower. If a piece of high-carbon steel and a piece of low-carbon steel are bound closely together and allowed to stand a long enough time, they will both have the same percentage of carbon. This would take many years at atmospheric temperatures, but if the steel is heated, each degree of temperature increase accelerates the flow of the carbon. When the molten stage is reached, minutes will produce, in the equalization of the carbon content, what it might take centuries to produce at atmospheric temperatures.

The principle here involved is that carbon flows to the body or element which has the greatest attraction for carbon. In carbonizing steel, a carbonaceous gas is produced, but the steel has a greater attraction for the carbon than the atmosphere in the carbonizing retort and hence it enters into a combination with the steel. Likewise, with two pieces of steel, the one low in carbon has a greater attraction than the piece high in carbon and draws this element away until the attractive force in each piece has been equalized. In manufacturing steel, the carbon is put into the molten metal and as the iron has a greater affinity for carbon than any of the elements in the slag, it combines with the iron and stays in the steel thus produced; hence, the correct percentage is found in the finished product.

We often find decarbonized spots in steel that has been heated for hardening in ordinary fuel-heated furnaces. A condition has here been produced whereby some element has entered the heating chamber that has a greater attraction than iron for the carbon; hence enough carbon flowed out of the steel to satisfy this attraction for the time this condition prevailed. Time and temperature seem to be the two factors that control the rate of flow of the carbon. Salt and lead-bath furnaces may produce these decarbonized spots. Sometimes a pitting of the surface occurs, which means that some element is present that has a great attraction for iron and has eaten it away.

In the electric furnaces previously referred to, no such elements seem to be present, as pitted and decarbonized surfaces are not found in the steels heat-treated in them. This applies to the carbon steels, which are heated to hardening temperatures as low as 1300 degrees F., and also to the high-speed steels that require a temperature as high as 2200 or 2300 degrees. Nearly all of the temperatures between these points are utilized, as the correct hardening temperature for some of the carbon steels is as high as 1500 degrees, while the alloy steels require from 1500 to 1700 degrees and some high-speed steels need a temperature as low as 1750 degrees. The elimination of pitting and decarbonizing is doubtless due to the fact that the oxygen in the air is not required to perfect combustion, and none of the products of combustion are present to attack the metal. As the steel leaves the furnace, no scale can be seen, because scale always comes from an excess of oxygen; neither is there any of the deposit that is sometimes produced by the products of combustion.

Effect of Oxygen on Iron and Steel. — Under certain conditions oxygen has a great affinity for iron and penetrates steel much more rapidly in a moist atmosphere, or one that has been heated. If steel is left in a damp place, it is only a question of time when it will be reduced to an iron oxide in a powdered form, the degree of dampness being the factor that governs the speed of this reduction. Thus, if steel is immersed in water it decomposes much more rapidly than when left in the air. Salt water reduces the time of this decomposition, as the salt aids the oxygen in forming its combination with iron. Without the aid of moisture, any increase in the temperature of steel increases the ability of oxygen to unite with the iron. When the boiling point is reached, the metal is saturated with oxygen and other gases, but the bulk of these are expelled when the steel solidifies. Enough is often left, however, to considerably weaken its various physical properties.

In furnaces that depend on combustion, or flames, for obtaining the necessary heat, a scale is liable to form on steel at the temperatures required for hardening, unless all of the oxygen is utilized to form the combustion and passes out of the vent as carbon dioxide (CO_2). This scale injures the piece being hardened, as it reduces its size and makes the surface uneven. This formation of scale proves that oxygen combines with steel at temperatures far below the melting point; hence the valves of such furnaces must be carefully adjusted so as not to deliver an excessive supply of air.

In addition to the formation of scale, oxygen injures steel in other ways. When the steel is heated to the higher temperatures, oxygen may enter the open pores to form microscopic bubbles and thus reduce the cohesive force that binds the molecules of the mass together. It may take the form of occluded gas in combination with other gases, or it may form a ferrous oxide with the iron. In any of these forms, oxygen reduces the strength, wearing qualities and resistance to fatigue or torsion stresses. Steels containing oxides also rust more quickly than those that are practically free from them. It is not so much the oxygen itself that is injurious, as it is the oxides that it forms with other elements.

The percentage of oxygen has heretofore been considered to be too small to be taken into account, but 0.05 per cent of oxygen is equal to 0.22 per cent of ferrous oxides, and this is sufficient to materially reduce the physical properties. As oxygen is a gas, it was difficult to analyze steels for this element, but with the new methods that have been devised, it has been found to be present in steel in larger quantities than was supposed, and efforts are now being made to reduce this impurity in all steels. That the pores of steel are opened by heat and allow the gases to enter or leave has been proved by a number of experiments where both cast and rolled steel have been heated in a vacuum. Under this condition, the gases began to leave the steel at temperatures between 300 and 600 degrees F. The volume reached a maximum at temperatures between 900 and 1000 degrees, and was then reduced to a minimum volume at about 1300 degrees. Another maximum point was reached at about 1450 degrees; then again reduced and again increased at higher temperatures. The greatest evolution of gas seemed to take place at the transformation point of the metal. That gases travel into the steel under atmospheric pressure is shown by the fact that oxygen combines with iron and raises blisters or scale and also that nitrogen penetrates steel that has been raised to temperatures around 2200 degrees.

Effect of Nitrogen on Steel. — Nitrogen is just beginning to be recognized as an injurious element in steel. It now occupies practically the same position that sulphur and phosphorus did but a short time ago, when the chemist first proved that they were very injurious elements to the physical properties and pointed out that means should be devised to keep them as low as possible. Then the practical steel makers scoffed at the chemists for saying that such small percentages of any element could weaken the steel and if it did their ancestors would have made it known. One does not have to be very old to remember the time when such talk was prevalent in the steel mills. It has been proved, however, that nitrogen is as liable to cause brittleness and "cold shortness" as is phosphorus, and is as injurious.

The better grades of steel contain from 0.005 to 0.025 per cent of nitrogen, while the cheaper grades contain between 0.010 and 0.065 per cent. Each increase in the percentage causes the elongation to diminish rapidly and the ductility to be reduced. At first, only a slight decrease occurs in the toughness, but the decrease becomes more rapid as the percentage of nitrogen increases. When the carbon content of steel is high, 0.035 per cent of nitrogen will cause the elongation and contraction to become practically nil, while in medium carbon steel it may take a nitrogen content of 0.050 per cent to accomplish this, and 0.065 per cent in low-carbon steels. Steels made in the resistance type of electric furnace contain only traces of nitrogen, but those made in the presence of basic slag in the arc type may contain an injurious amount.

As four-fifths of the air is composed of nitrogen, by volume, consuming the oxygen with a flame would leave much of the nitrogen free to enter the pores of steel that is heated to the hardening temperatures in ordinary oven furnaces, especially when the steel is heated to the 2200 degrees F. or more required for hardening some brands of high-speed steel. Decarbonization is doubtless due to the nitrogen that may be occluded in steel combining with carbon to form cyanogen, which escapes when the metal is heated. As the resistance type of electric furnaces does not consume any of the oxygen in the air, none of the nitrogen is set free. Thus a condition that allows the nitrogen to penetrate the steel is not created, even though the steel be heated and the pores open. As steel pieces do not warp as much in this type of electric furnaces, as in coal-, oil- or gas-fired furnaces, it may be because the metal is more dense, owing to its not absorbing any gases.

The combustion type of furnace, especially when using coal as fuel, gives off many elements as products of combustion which may be injurious to steel. Among these may be mentioned such hydrocarbons as anthracine, naphthalene, toluene, benzine, methane, ethane, etc., while acetylene, benzole and sulphur are other products sometimes present. With a large amount of hydrocarbons present in a furnace, it is almost impossible to prevent carbonization in the steels, as the gases that are commonly present, other than nitrogen, have no other effect on steels. Thus, some parts of the piece being hardened would be carbonized more than others and cause a variation in the hardness. To prevent this, the muffle type of furnace should be used. The instruction given with all high-grade steels is to heat the steel in a muffle furnace so that the products of combustion, or the flame, cannot attack the metal. The construction of the electric furnace, however, is such that it forms its own muffle and the expense of renewing muffles is done away with.

Current Consumption and Operating Cost. — A resistance furnace of the type described, with a heating chamber 12 inches wide, 18 inches deep and 8 inches high, can be brought up to the higher temperatures required for high-speed steel, with 30 kilowatts, in one hour and fifteen minutes, and then be maintained at this temperature with about one-half of this current. With an average amount of steel to heat, it can be operated during a ten-hour day with about 145 kilowatts. The heat insulation is so effective that furnaces have been closed after turning off the current, and at the end of twelve hours the temperature had only dropped to between 700 and 800 degrees. Smaller furnaces with a heating chamber 7 inches wide, 12 inches deep and 5 inches high, only use one-half the current mentioned.

The current cost for an electric furnace is undoubtedly higher than the fuel cost for a gas-, oil- or coal-fired furnace of the same size, but when all other things have been taken into consideration, this current cost is minimized and the electric furnace can be made to compete commercially with other kinds. This is especially true when quality is a factor in the work heat-treated. The ease with which the temperature can be controlled and accurately maintained; the absence of scale or pitting; the decrease in warpage; the cleanliness of the work due to the absence of deposits of any kind; the overcoming of decarbonized spots; the lowering of the penetration of gases or other injurious elements, such as sulphur; and the uniformity in the hardness or temper of the steel, are all things which should be figured against the cost of fuel. In addition to this, the electric furnace can be so heat-insulated that the hand can be held on the outer shell or case without burning. This lowers the temperature of the hardening room and makes the working conditions so much better that operators are enabled to turn out more work. One company cites a case where three tons of cold-rolled steel were heat-treated per week, in the shape of razor blades, in six electrically-heated furnaces with two operators; whereas, before the installation of electric furnaces, it required fifteen men, and sixty-five furnaces that used gas and blast, for the hardening operations on the same tonnage. While this might be a case where the work was particularly adapted to electric furnaces, many others can doubtless be found, where a saving could be effected when all of the factors are figured into the cost.

A report from the Timken-Detroit Axle Co. relating to the results obtained with the furnaces in their plant states that with the four furnaces mentioned each hour of the ten-hour work day averages 75 pounds of steel gears heated to the hardening temperature in each furnace. The average current consumption is 13 K. W. per hour for each furnace, costing 1¼ cent per K. W. The average cost per pound of steel hardened is a trifle more than 0.2 cent.

Chapter Seven - The Metallic Salt-Bath Electric Hardening Furnace

In externally-fired furnaces, the heat losses are always considerable, and only a small part of the energy used in heating is utilized for raising the temperature of the metal to be hardened. There is also a disadvantage in employing gas- or oil-fired furnaces in that the high temperatures rapidly destroy the crucibles. Electric hardening furnaces, therefore, possess marked advantages for this work over the various types of externally-fired furnaces.

The electric furnace described in the following has been brought out by the Allgemeine Elektricitäts-Gesellschaft of Berlin, Germany, and in this country by the General Electric Co., Schenectady, N. Y. Briefly described, the furnace consists of a bath of melted metallic salts contained within a firebrick crucible, inside of which, at two opposite sides, are fixed electrodes of iron very low in carbon, the melting point of which is higher than that of ordinary steel. This crucible is surrounded by a thick layer of asbestos, which is, in turn, imbedded in a layer of some heat-insulating material, the whole being held together by a steel case. The walls of the furnace are made so thick in relation to the dimensions of the crucible that the steel case of the apparatus may be touched with the hand without injury after having been in operation for hours, at the highest temperatures required in the hardening room.

The soft iron supply conductors to the electrodes are connected to the secondary copper bars of a regulating transformer which transforms the normal voltage to the low voltage (5 to 70 volts) employed in the operation of the furnace. A topical arrangement of the equipment of a large works has the furnaces provided with a hood in a central position, and a quenching tank immediately beside the furnace on one side. By this latter arrangement the change in temperature caused by carrying pieces from the furnace to the water tank is reduced to a minimum. The tank is supplied with heating and cooling coils with steam or cold water, so that the temperature of the quenching bath can be easily regulated.

Requirements of Hardening Furnace. — A great many factors must be considered in the development of the design of an electrical hardening furnace. The practical requirements which should be fulfilled by an ideal hardening furnace may be summarized in a general way as follows:

1. The furnace should make it possible to obtain all hardening temperatures required in industrial practice, thus having a range of from 1400 to 2450 degrees F.

2. The steel should be heated to the required temperature easily and rapidly.

3. The temperature of the steel should be easily ascertained, and it should "be possible to keep it well under control within a margin of, say, 50 degrees F. above or below the exact temperature required.

4. The steel must be equally heated all over, notwithstanding different cross-sections of the object, thus preventing the overheating and burning of edges and points.

5. During the heating process foreign matter must not come in contact with the steel so as to change its carbon content, or affect it in other respects.

6. It should be possible to place the cooling tank close to the furnace in order to minimize the loss of heat during the transfer, and avoid the oxidizing influence of the air.

7. The furnace should not give off obnoxious or poisonous vapors of lead, potassium-cyanide, etc.

8. The total operating cost incident to the hardening process should be low.

The electric hardening furnace described in this chapter is claimed to fulfill to a considerable extent all of the previous requirements, and a general review of the advantages of electric hardening will be given. It should be understood, however, that many of the claims made at the introduction of this furnace have not been substantiated by experience, as will be explained in the latter part of this chapter.

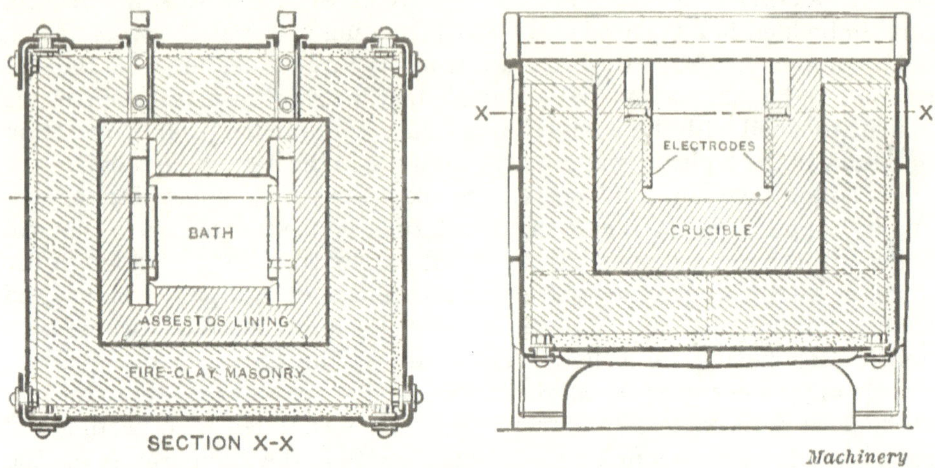

Fig. 1. Horizontal and Vertical Sections through a Metallic Salt-bath Electric Hardening Furnace

Description of Hardening Furnace. — In Fig. 1 are shown vertical and horizontal sections of the hardening furnace. A bath of metal salts is contained in a fireclay crucible. Current is transmitted .to the bath by two electrodes made of Swedish ingot iron, which is characterized by a particularly low percentage of carbon, and therefore has a melting point of as high as 2700 to 2900 degrees F. As shown in the horizontal cross-section, the electrodes end in iron terminals sweated in turn to copper conductors. The crucible is surrounded by an asbestos lining, a fireclay receiver, and a layer of insulating material, the whole being contained in a cast-iron case. This construction greatly reduces the radiation losses, and after ten hours' operation of the furnace at about 2450 degrees F., the cast-iron case has a temperature of only from 85 to 105 degrees F. Over the bath a sheet-iron hood is placed fitted with chimney and damper. These furnaces are made in several different sizes. In the smallest size the inside dimensions of the crucible are about 5 inches square by 5 inches deep, and in the largest size 12 inches square by 15 inches deep. The approximate consumption of current in kilowatts at various temperatures for the large and small furnaces is given in Table 1.

The best composition of the bath depends mainly on the temperature required for the hardening. Table 2 gives the composition of various salts to be used for different processes. The conductivity of the salts at normal temperature is very small, while at high temperatures (when in a melted condition) they offer to the electric current a comparatively low resistance. When the mixture is sufficiently hot, the bath, therefore, forms an electric conductor, and each part of the bath produces its own heat. This feature distinguishes this class of electric furnaces from other types.

Temperature of Bath, Degrees F.	Size of Crucible of Furnace, Inches			
	5 × 5 × 5	6 × 6 × 7	8 × 8 × 11	12 × 12 × 15
1400	2.5	3.5	7.5	17.5
1550	3.0	4.5	8.5	20.0
2100	5.5	9.0	16.0	36.0
2350	7.5	12.0	22.0	48.0

Table 1. Current Consumption of Electric Furnaces in Kilowatts

Process	Temperature, Degrees F.	Salts
Tempering steel.........	400 to 1075	Sodium nitrate and potassium nitrate
Annealing copper, alloys, etc..............	1200 to 1650	Sodium chloride or sodium chloride and potassium chloride
Hardening regular carbon steel.............	1400 to 2000	Potassium chloride and barium chloride
Hardening high-speed steels...............	1900 to 2450	Barium chloride
	2700 to 2900	Calcium fluoride or magnesium fluoride

Table 2. Temperature and Composition of Heating Bath

The heating of the salts prior to their becoming highly conductive is done by means of an auxiliary electrode and a piece of arc lamp carbon. The carbon is first pressed against one of the main electrodes and soon reaches a white glow, melting the salts immediately about it. The auxiliary electrode, which consists of an iron stick fitted into a wooden handle, is then drawn towards the other main electrode, the molten salt trailing behind it until a bridge is established between the two main electrodes. The current which now passes through the molten salt continues to raise the temperature of the bath until the required heat is attained. The articles to be heated are dipped into the bath, suspended by thin iron wires or held by tongs, and are allowed to remain in the bath until uniformly heated throughout.

The most striking feature of this furnace is the possibility of securing uniformity in temperature throughout the whole bath. Careful measurements with a pyrometer of the thermo-couple type at various parts of the bath have shown that the temperature varies only 5 or 6 degrees F., except in an upper layer about ½ inch thick, where, owing to radiation, the temperature is from

20 to 35 degrees F. lower. Alternating and not continuous current should be used; all frequencies between 25 and 60 cycles may be applied; with less than 25 cycles electrolytic phenomena appear. The furnace, having only two electrodes, is suitable for single-phase currents only. If a single-phase supply is not available, a converter must be installed.

An important part of the hardening installation for electric furnaces is the pyrometer. The most reliable results for the temperatures in question are obtained by instruments of the thermocouple type, but the instruments must be such that the terminals are kept outside of the destructive influence of the heat. The thermo-couple used is platinum — platinum-rhodium, protected by Marquardt composition and steel. A steel cylinder protects the parts projecting from the bath. This cylinder gets white hot and deteriorates unless protected against the oxidizing influence of the air.

When the salts are melted, the voltage necessary for maintaining the temperature is from 5 to 30 volts, while the heating-up voltage is about 70 volts. Such low voltages are not available from ordinary supply systems, and consequently a transformer must be used. The heat developed, and consequently the temperature of the bath, depends on the voltage. If it is desired to alter the temperature, this can therefore be done by a variation of the voltage. The use of the transformer makes the voltage control comparatively simple.

The Hardening Process. — When heating carbon steel for hardening, it is advisable to heat it as rapidly as possible, because the prolonged influence of the heat seems to affect the chemical constitution and mechanical structure. In most cases it is advisable to pre-heat the steel to a certain temperature before the final heating in the bath takes place. This pre-heating should be done thoroughly in order to make sure that all portions are well heated. Unless this is done during the pre-heating period the heat during the main heating period must be directed to the parts not properly heated, which lengthens the process and may cause damage to the external portions of the tool, such as edges, projections, etc.

Each brand of steel requires a certain temperature to which it should be heated for hardening. For high-speed steels this temperature is from about 1800 to 2375 degrees F., and for carbon steels from about 1300 to 1650 degrees F. Generally speaking, the cooling process for alloy steels need not be so abrupt as for carbon steels. Instead of quenching the hot tool in water or oil, it is sufficient to expose it to a current of air or to dip it into molten tallow.

Advantages of Electric Furnaces. — A great advantage of the electric furnace is that it is possible to cover a wide range of temperatures with one equipment by only changing the composition of the bath. The tool can thus stay in the bath until it has acquired the temperature of the bath, and does not need to be removed before it has assumed the temperature of its surroundings, which is quite commonly the case in other heating processes. With the electric hardening bath, having a predetermined temperature, less dependence is placed upon the skill of the operator, and no account need be taken of the fact that smaller cross-sections heat up quicker than larger ones.

When dipping cold steel into the heating chamber, the temperature of the latter must drop. In fact, with gas-fired furnaces and salt baths it falls rapidly, unless the uncertain procedure of increasing the gas supply is resorted to. In the electric furnace, when the tool is dipped into the salt, the level of the salt bath rises, and the current of heat produced increases automatically. Besides, when it is necessary to immerse large solid masses, the current supply can be easily increased by the regulator, thus preventing a drop in temperature.

Of course, the smaller cross-sections of the tool will heat up quicker than the larger ones in the electric furnace as well as in other heating furnaces, but the delicate parts will not *overheat* because they cannot assume a higher temperature than that of the bath itself. The bath equalizes all differences in temperature, and in a very short time heats the whole mass uniformly. This explains the very small loss from overheating in electric furnace plants as compared with others.

While the tool is in the bath, the air is, of course, prevented from coming in contact with it, but a thin coating of salt protects it still further when on its way from the bath to the cooling tank, and falls away first when the object is placed in the cooling liquid. This is a great advantage over all types of open-fire or muffle furnaces, but is common to all bath-type furnaces. Metal salts, moreover, offer the advantage that they do not give off poisonous gases, and unlike lead, they can be obtained comparatively pure at a reasonable cost. The salt coating also breaks up entirely in the cooling liquid, while when tools are heated in lead, small particles of it sometimes stick to the steel, leaving soft spots on the hardened surface.

During the heating-up period or when a certain temperature is exceeded, the melted salts give off a small amount of vapor, and therefore a hood and chimney are provided for the furnace, but during normal operation there are scarcely any vapors produced. The hood offers the further advantage that the radiation from the bath surface can be used for the pre-heating of the articles to be hardened. A grate may be fixed in the hod in which the articles are placed, prior to being dipped in the bath.

Comparative Operating Cost. — The parts subject to wear in an electric furnace are the crucible and electrodes. The crucible has been found to have a life of from 1200 to 1800 hours at a temperature of 2350 degrees F., and up to 3000 hours at lower temperatures. This is much longer than with muffle furnaces, which is probably due to the absence of the destructive influence of the gases of combustion, and to the fact that the crucible does not transmit the heat from the outside to the inside. The most sensitive part of the electrodes is that which projects over the level of the bath, and which is protected by exchangeable tips. These tips have a life of from 400 to 800 hours, and the cost of their replacement is as low as fireclay for other furnaces. The amount of salts lost by evaporation and waste under ordinary working conditions in a furnace 8 X 8 X 11 inches amounts to a little more than one pound for ten hours' continuous operation.

The ease with which the electric furnace can be handled makes it possible to use cheaper labor than that employed in plants where the success of the work depends on the skill of the operator. The speed of the hardening process is also much greater, and therefore a larger number of pieces can be handled per hour.

Even at a temperature of 2400 degrees F., attainable in laboratory tests, but not usually employed in commercial hardening, the damage to the crucibles of the electric furnace is very small. Working ten hours a day with this temperature, a crucible will last six months, and for ordinary hardening temperatures, fifteen months.

Results obtained. — By means of this process, it has been possible to harden large high-speed steel milling cutters in about half an hour, including the time for pre-heating, which takes the greatest part of the time. Bringing the cutters up to a temperature of 750 degrees F. constitutes this preheating. After that, it takes only about a minute to bring an average-sized cutter to 1400 or 1500 degrees F., and then another minute to bring it up to about 2370 degrees F., which is considered the right hardening temperature for some brands of high-speed steel. At this temperature, however, the barium chloride attacks the steel, as will be referred to in detail on the following pages. The time stated above refers to average-sized and heavy milling cutters, whereas it only takes from 6 to 10 minutes to bring a small milling cutter to the right temperature in the electrically heated salt bath.

In regard to cooling the cutters, it has been found that when high-speed steel tools are cooled in an air blast, any moisture coming in contact with the hot tool has a tendency to crack it, so that it becomes necessary to dry the air before it enters into the nozzles. It has also been found that it is absolutely impossible to cool a cutter which has a very heavy body and fine teeth in the air blast, as the heat from the central portion is not extracted fast enough, and therefore does not permit a sufficiently rapid cooling of the teeth to insure proper hardening. For this reason, some firms have adopted a method of cooling the cutters from the hardening heat of 2370 degrees F. to a temperature of about 1100 degrees F. by quenching in an electrically heated salt bath. After having been cooled to about 1100 degrees F. in the bath, the cutters are allowed to cool down slowly in the air, and the whole process has the advantage of being cheap and reliable as well as effecting a considerable saving in time.

It must, however, be understood that electrically heated barium salt baths are advantageous to use only when a large quantity of tools is to be hardened, because this method will otherwise prove expensive. It has also been remarked that the electrically heated bath is more advantageous for heavy than for small tools, but it is not clear why the process should be thus limited to the former class of tools. The disadvantages of the barium-chloride bath will now be taken up.

The Use of Barium Chloride for Heating Steel for Hardening. — As is well known, high-speed steel requires to be heated to a much higher temperature for hardening than does ordinary carbon steel. While a heat of from 1400 to 1600 degrees F. is sufficient for tools made from carbon steel, a heat of from, at least, 1800 to 2200 degrees F. is required in order to satisfactorily harden high-speed steel tools. The ordinary lead bath commonly used for heating carbon steel tools cannot be used at such high temperatures as these, and as it is, in general, unsatisfactory to heat the tools in an oven furnace, owing to the difficulty of correctly determining the hardening temperature when the tools are heated in this way, some heating medium has been sought which could stand high temperatures and in which the pieces to be hardened could be immersed so as to obtain a uniform heat without danger of burning delicate points or cutting edges — a danger which is always present when high-speed steel tools are heated to a high temperature in an open heating furnace. It has been believed that a satisfactory heating medium had been found in barium chloride, and this medium has been, and is still, used to a considerable extent both in this country and abroad; but the results obtained have not been as favorable as was at first expected, and many users of barium chloride have abandoned its use on account of the difficulties met with.

It appears that tools heated for hardening in a crucible containing barium chloride have a soft scale or film of soft metal, probably 0.003 to 0.006 inch deep, all over the surface of the tool. Careful experiments have been made to ascertain as nearly as possible the conditions which contribute to produce such unsatisfactory results. Comparison has been made between tools made from the same material of which some were hardened by heating them in barium chloride and some in an oven furnace. The results of these experiments are recorded in the following.

In order to make the tests as simple, and at the same time as conclusive, as possible, pieces of high-speed steel, ½ inch thick, were cut off from one bar of steel. These pieces were hardened by heating some of them in a common oven furnace, and others in barium chloride melted in a graphite crucible placed in a gas furnace. The pieces were heated directly from the room temperature to the hardening temperature, no pre-heating being resorted to. The barium chloride used was chemically pure. The temperatures were recorded by a Bristol pyrometer, and the hardness tests were made on a Shore scleroscope. After heating, the pieces were immersed in a cooling bath consisting of cottonseed oil at a temperature of 100 degrees F. The temper was then drawn in an oil tempering bath at 500 degrees F., a temperature which is not too high for the higher grades of high-speed steel, although it would be excessive for ordinary carbon steel.

When the pieces were heated in the oven furnace, the operator, an experienced hardener of this kind of steel, used his own judgment as to when to

Table Showing the Hardness of High-speed Steel Heated for Hardening under Different Conditions

Heat of Hardening Bath, Degrees F.	Heated in Oven Furnace			Heated in Barium Chloride			
	Time Test Piece was Left in Furnace, Minutes	Degree of Hardness on Scleroscope Scale		Left in Bath 10 Minutes		Left in Bath 18 Minutes	
				Degree of Hardness on Scleroscope Scale		Degree of Hardness on Scleroscope Scale	
		Face of Test Piece	Back of Test Piece	Face of Test Piece	Back of Test Piece	Face of Test Piece	Back of Test Piece
1700*	9½	Soft	Soft	Soft	Soft	Soft	Soft
1800	7	83.5	85	92	91	93	90
1900	6	91	86	93	91	91	92
2000	6	90	86	91	87	92	89
2100	5½	90	91	91	82	87	74
2200	5	93	91	88	82	86	70
2300	4¾ †	93	93	73	64	74	62
2400	3½	92	93	65	65	63	57

[*] At this temperature the steel would not harden, and, therefore, no scleroscope tests were made.

[†] This sample was burnt and pitted, indicating that it had been kept in the fire too long.

remove the piece from the furnace and plunge it into the hardening bath, but the time required for the piece to acquire proper hardening heat was recorded, and is given in the accompanying table. The degree of heat as given, is the heat of the furnace as recorded by the pyrometer, but it is evident that in the case of a piece of steel heated in an oven furnace and removed according to the judgment of the operator, there may be a slight variation between the heat of the furnace and the heat of the piece itself. When the tools are heated in the barium chloride bath, the temperature of the piece and the bath will, of course, be the same, provided the piece is permitted to remain in the bath long enough, which was the case in the experiments described.

After the pieces had been hardened and tempered as described, an amount equal to 0.005 inch was ground off from one side of the pieces, which we will call the face, and an amount of 0.002 inch was ground off from the other side, the back. The surfaces presented to the scleroscope were thus perfectly smooth and uniform, but it should be noted that less of the soft scale, mentioned in the foregoing, was removed from the back of the pieces than from the face. The pieces were now subjected to scleroscopic tests, carefully recorded and repeated several times. The results of these tests are given in the accompanying table, the values given being the average of the several readings.

It should also be mentioned that the pieces heated in the barium chloride at 2100 to 2400 degrees F. were found to be pitted, and small beads of a metallic structure adhered to the pieces. Similar small pieces were found in the bottom of the crucible after all the test pieces had been hardened. This residue was chemically analyzed and was found to consist principally of ferro-tungsten, the analysis showing tungsten, iron and carbon to be present. The carbon content was about 3.3 per cent, tungsten 9.8 per cent, and iron 86.9 per cent.

Several interesting and instructive conclusions with relation to the heating of high-speed steel in an oven furnace, and the action of barium chloride as a heating medium for high-speed steel when hardening, may be drawn from the results recorded in the table. It will be seen that when heating in an oven furnace, the results obtained were almost uniformly better according to the heat at which the pieces were hardened. The higher the heat, the higher the scleroscopic test number. This result is in thorough harmony with the general principle that the higher the heat at which high-speed steel tools are hardened, the better their cutting and "standing up" qualities. When the pieces were heated in barium chloride, however, a result entirely different was obtained, and at temperatures of 2100 to 2400 degrees F., the results were, in general, very unsatisfactory. In the case where the pieces were permitted to remain 18 minutes in the heating bath it will be seen that the face of the piece is almost uniformly softer, the higher the hardening heat. This may be taken to indicate that there still was some of the soft scale left, even after having removed an amount equal to 0.005 inch by grinding. A file test on the surface, however, could not detect this scale, as the surface seemed glass-hard.

The feature which will particularly be noticed in studying the table is that in almost every case the back, where only an amount of 0.002 inch was removed, is softer than the face of the test piece. It is evident that this 'is due to the fact that the soft scale is deeper than 0.002 inch, and has not been entirely removed by the grinding on the back; whereas the face, where an amount of 0.005 inch has been ground off, is practically freed from the soft scale, and hence shows a greater hardness when tested by the scleroscope. The influence of this soft film is especially apparent when the steel is hardened at a temperature of 2100 to 2400 degrees F.

Having ascertained through the tests mentioned that barium chloride had a detrimental influence upon the hardness of high-speed steel heated in it at high heats (2100 to 2400 degrees F.), tests were next made to ascertain the influence on the cutting qualities of tools hardened either by heating in barium chloride or in an oven furnace. These tests proved conclusively that the tools heated in the barium chloride bath did not stand as high a cutting speed as did those hardened by heating in an oven furnace. The ferro-tungsten found in the bottom of the crucible indicates that, particularly at high heats, some of the tungsten and carbon is removed from the tools into the bath, thus changing the structure of the surface of the tool being heated. When an amount of, say, 0.010 inch is ground off from the cutting edges of tools, the influence of the heating in barium chloride is less noticeable — in fact, sometimes not noticeable at all — but when the tools cannot be ground after hardening, barium chloride is not a heating medium which can be recommended under any circumstances. The change of the structure on the surface of the tool explains why tools heated in barium chloride cannot stand up at as high speeds as those heated in an open fire.

Another disadvantage met with in the use of barium chloride is that the residue of ferro-tungsten found in the bottom of the crucible seems to have a deteriorating influence on the crucible, "eating" through it in a comparatively short space of time. As a general conclusion it may be stated that whenever barium chloride is used as a heating bath, it should never be permitted to reach a temperature of more than 2050 degrees F.

Barium Chloride in the Electric Hardening Furnace. — The difficulties met with in the use of barium chloride in a crucible heated in a gas furnace are still further accentuated when using an electric hardening furnace of the type employing a barium chloride bath as the heating medium.

When steel is heated in barium chloride in an electric furnace, the current apparently passes directly through the steel and the pieces are heated not only by the heat imparted to them from the barium chloride bath, but also by the resistance to the electric current passing through the steel itself. That this must be the case is indicated by the fact that tools heated in an electric furnace are brought up to the proper temperature for hardening in approximately one-third of the time which is required for heating them in a barium chloride bath of the same temperature contained in a graphite crucible heated in a gas furnace. As an example it may be mentioned that certain tools which must remain sixteen minutes in a barium chloride bath in a gas furnace can be heated in the bath in the electric furnace in from four to five minutes. The barium chloride bath in an electric furnace does not cool down to the same extent when the pieces to be hardened are immersed, as does the heating bath in the gas furnace. This also indicates that in the electric furnace the heat required for bringing the steel to a hardening temperature is only partly derived directly from the bath in the electric furnace; while all of the heat required must be given out by the barium chloride bath in the crucible in the gas furnace. These facts make it conclusive that there is an entirely different action in the heating of steel immersed in a bath of the same character in an electric furnace than there is when it is immersed in a bath contained in a graphite crucible. The rapidity with which tools to be hardened can be heated in an electric furnace has been quoted as one of its principal advantages, and so it would be were it not for the fact that the surface of the steel deteriorates under the action of the bath, the bath in turn deteriorating under the action of the electric current.

On tools which are ground after hardening, there does not seem to be any difference between the hardness of those which have been hardened in barium chloride in a crucible and those heated in the same medium in the electric furnace. The tools can be run at the same cutting speed and will stand up equally well; but when the tools cannot be ground on the cutting edges, as in the case of formed milling cutters, knurls, taps, dies, etc., then the tools heated in the electric furnace are decidedly inferior, especially under certain conditions which will be more thoroughly explained in the following, there being a thin scale or film on the outside which is entirely too soft to possess proper cutting qualities.

Experiments have been undertaken in order to determine, to some extent at least, the causes of the difficulties met with in heating tools in the electric hardening furnace. No conclusive answers can, perhaps, be given to all of the questions which may be asked in this connection, but the results of the experiments give at least a clue to the cause of the trouble, and further experiments might be made which would give still more conclusive results; if methods can be developed which will remedy the defects and make it possible to heat steel in a barium chloride bath in an electric furnace without having to contend with the soft scale on the surface of the steel, the electric furnace would present the best means for heating tools in a bath of high temperature, on account of the decided difference in the time required to bring the tools to the proper hardening temperature.

It has been found that barium chloride when used in a graphite crucible in a gas furnace slightly deteriorates when it has been used for a number of days; but the difference in the results obtained when heating in a bath of entirely new barium chloride and a bath which has been in use for several days, say from six to ten days, is so small that ordinarily no attention need be paid to it. Some users of the barium chloride bath, however, have been in the habit of changing the bath every day, using new barium chloride at all times. This practice is, of course, very expensive, and the advantages gained are too small to warrant the added cost. When barium chloride is used in an electric furnace, however, it deteriorates very rapidly, so that it is practically useless for its purpose after a couple of days' use, and after having been used for a week it may be stated without exaggeration that it is entirely unsuited for any further use. Furthermore, the barium chloride which has been used in a graphite crucible generally has an almost white color after several days of use, whereas that used in the electric furnace is of a dark gray color. This difference in color is apparently due to the fact that the barium chloride in the electric furnace dissolves the ferrotungsten which, as previously mentioned, is found in the bottom of the crucible when heating steel in a gas furnace; or possibly the soft iron electrodes are partly dissolved by the barium chloride. The absence of a precipitate in the electric hardening furnace and the gray color of the bath would seem to make it safe to draw this conclusion.

As regards the deterioration of the barium chloride after a few days' use in the electric furnace, several interesting facts have been noted. When steel is hardened after having been heated in a bath consisting of new barium chloride which has been used but one or two days, it seems to acquire a satisfactory hardness, at least as satisfactory as when heated in the same kind of a bath in a crucible heated in a gas furnace. There is, of course, a very thin soft scale, the same as on all tools heated in a barium chloride bath, but this scale is so thin that it cannot be detected with a file test, and hence can be considered of no consequence. After the first day or two, the influence of the barium chloride in conjunction with the electric current on the surface of the steel is considerably augmented; and when the barium chloride has been in use in an electric furnace for about a week, a scale from 0.005 to 0.010 inch deep can be easily detected. A scale of this thickness, of course, makes it impossible to use this means for heating the steel in any case where the cutting edges of the tools cannot be ground off to a sufficient depth to entirely remove the soft outside portion. In the experiments made, the steel was heated to a temperature of about 2100 degrees F.

It was thought that possibly some other factors besides the barium chloride, which had been in use for a number of days, were the cause of the soft scale found on the tools. New barium chloride was, therefore, melted in the furnace, and a new set of tests made. In these tests the results obtained in the first series were entirely duplicated. During the first two days the steel hardened had no perceptible soft film or scale; but when the bath had been in use for about three or four days there was a pronounced soft scale, and after a week, the scale had a thickness practically the same as in the first series of tests.

While, as already mentioned, no indications of the presence of ferro-tungsten were found in the bottom of the electric furnace, a soft, dark gray precipitate was found in the bottom of the electric furnace pot when it had been run with the same bath for about a week. The electrodes of the electric furnace are made of soft iron, and it is likely that the precipitate found originates from the electrodes, as it proved to be a metallic substance similar to iron, and soft enough so that it could be easily filed, which, of course, would not have been the case had the precipitate consisted of ferro-tungsten.

Another interesting fact has also been noticed in connection with the electric furnace. The amount of barium chloride used in a given time is much greater in an electric furnace when the bath is heated to a given temperature, than it is in a gas furnace with a bath at the same temperature. For some reason the electric current passing through the bath seems to make it easier for the molten salt to volatilize.

The makers of electric hardening furnaces may undertake experiments which will give more complete data relating to the action of barium chloride in an electric furnace than those brought forth in the foregoing. At present, however, it seems that barium chloride is unsatisfactory for hardening high-speed steel in an electric furnace, and that it has only a comparatively limited application when used in crucibles and heated in a gas furnace. Some users of barium chloride baths for heating tools for hardening have, as mentioned, entirely abandoned its use after several attempts to make it successful, while others still continue its use for parts and tools on which a thin soft scale is not objectionable and can be removed by grinding.

It should be noted that in speaking of an electric heating furnace, only those furnaces using a barium chloride bath for the heating medium have been referred to. Other electric heating furnaces are made, in which no heating bath is used, but the steel is heated in the air between two electrodes. This type of electric heating furnace is in a class by itself, and while it presents some difficulties, they are of a minor nature and do not come under the head of the present investigation.

It should also be understood that the electric hardening furnace, using a barium chloride bath, may have a field of usefulness for heating ordinary carbon steel tools for hardening, as in this case hardly any of the objections mentioned in the foregoing, and which apply to the hardening of high-speed steel only, are present, except the increased consumption of barium chloride and potassium chloride, with which latter salt the barium chloride must be mixed when used for heating carbon steel. This mixture is necessary in order to obtain the lower melting temperature of the bath required for carbon steel. The advantages of the electric hardening furnace for carbon steel tools are the greater rapidity with which they can be heated and the clean white surface on the tools thus hardened. The cutting edges of the tools seem to be protected by the coating from the heating bath which falls off when the tools are dipped. When dipping in an oil bath, this coating is not entirely removed, but it always disappears when dipping in hot water or soda solution.

Chapter Eight - Miscellaneous Types of Electric Furnaces in General Use

The growing demand for a higher quality in heat-treated steels and for a constant improvement of the physical properties of the metal is making the electric furnace a commercial necessity, both in the manufacture of the metal and in its hardening and tempering. Efforts are also being made to use electric furnaces in the fabrication of steels, and such furnaces are now in use for heating steel before it is drop-forged and bent to shape for leaf springs. This is due to the fact that when so heated impurities which are injurious to steel are either removed from or not allowed to penetrate the metal, and these impurities are thus present in lower percentages than when the steel is heated in furnaces that use a flame to generate the heat. Another reason is the ease with which the temperature can be controlled and, consequently, the greater accuracy that can be obtained in securing the correct degree of temperature that is required for heat-treating or refining the steel. This has caused many different types of electric furnaces to be developed for heat-treating steels and several different principles are utilized in their design and construction. Of these, the Hoskins type was described in Chapter Six, and consequently will not be considered here. Most of the remaining types use either the arc or resistance principle for supplying the necessary heat to the oven or liquid bath furnace.

A recently developed type of electric furnace is the Baily furnace, built by the Electric Furnace Co. of America, Alliance, Ohio. In one plant where this furnace is installed it is used in the manufacture of leaf springs. In this instance the operation consists of inserting a leaf in the furnace and bringing it up to the fabricating heat, which must be 1800 degrees F. to make it bend properly to shape. The leaf is then removed and bent, after which it is replaced in the furnace, brought up to the hardening temperature and then taken out to be quenched in oil contained in a tank. After that, it is again placed in the furnace where the oil is burned off to get the drawing temperature.

It can be seen at a glance that this is a misuse of the electric furnace, as it eliminates its greatest point of excellence, namely, the accuracy with which the temperature can be controlled. The correct hardening temperature of carbon spring steels is about 1500 degrees F. To reinsert the leaves in a furnace, the temperature of which is maintained at 1800 degrees, or more, and get them all within 1 50 degrees of the correct hardening temperature, during a day's run, is an impossibility. Occasionally one leaf may be taken out at the right temperature, but no means have been devised for telling when the whole length of a leaf has reached a temperature of 1500 degrees F., when it is in a furnace heated to 1800 degrees. Man's eyesight is not sufficiently delicate or accurate. Burning the oil off to obtain the drawing temperature is not to be spoken of seriously. If a separate furnace were used for the hardening operation, the heat in the oven could be maintained at the correct temperature, which would be 1500 degrees in this case. The steel could be allowed to remain in the oven of the furnace until it had reached a uniform temperature and then be taken out for quenching. Pyrometers could be used to measure the temperature of the furnace oven, which would also be the temperature of the steel, and a more uniform product would be obtained. Another furnace operated at the correct drawing temperature would insure uniformity for that operation. The springs would then be much stronger and more resilient. The work could be turned out much faster in this way and with cheaper labor, which would lessen the cost of production.

The furnace is supplied with all the accessories that are necessary for obtaining accurate temperatures and maintaining them during any period of operation. With these accessories the voltage can be varied to control the number of kilowatts of power supplied to the furnace, thus regulating its temperature. The switchboard holds an ammeter, or current indicator, to show the amount of current that is flowing at any given time; a wattmeter to show the amount of power that is consumed per hour, day or month; a voltmeter to show the voltage of the current; and a recording pyrometer that shows on the chart the temperature the furnace has attained at any time during the day. These charts can be filed away to keep a permanent record of the temperature of the furnace on any work that is turned out.

No flames, smoke, gases, fumes or roar comes from this furnace; therefore, no chimneys, piping or blast are needed and an air blast or steam is not required, as in the case of furnaces that use a flame. The temperature of the room is also lowered, as most of the heat is confined inside the furnace. Thus, the working conditions are made so much more pleasant that the operators are able to produce more springs for a day's work.

This furnace is called a loo-kilowatt type, which means that its maximum consumption of power is 100 kilowatts per hour. The heating chamber is 5 feet wide by 7 feet deep by 1½ foot high and 600 pounds of steel plates can be heated to 1650 degrees F. in one hour. With all openings closed, the furnace can be maintained at this temperature with about 20 kilowatts of power, which replaces the heat lost through the walls, this result being obtained by adjusting the switch so that 200 amperes at 100 volts will be flowing into the furnace. It is based on the theory that each kilowatt liberates 3412 B.T.U. in the furnace. If the doors of the furnace are left open, it will require another 8 kilowatts to replace the heat lost through them, and if there are any holes in the sides or roof of the heating chamber for the heat to leak out, it would require considerably more power to keep the furnace up to the desired temperature. Therefore, it is important that no openings should be allowed to remain in the walls, and as the furnace can be completely relined with silicon-carbide, at a cost of some $2 or $3, there is no excuse for leakages to occur.

With the preceding data, it is a simple matter to calculate the amount of current required to heat a given amount of steel. To heat one pound of steel to 1650 degrees F. requires 300 B.T.U.; therefore to heat 600 pounds every hour would require 180,000 B.T.U. per hour. This result divided by 3412 gives 53 as the number of kilowatts per hour that the furnace requires to generate this amount of heat. If we add 20 kilowatts for the wall loss and 8 kilowatts for the loss through the door, we have a total of 81 kilowatts per hour as the amount of power that is needed to heat 600 pounds of steel per hour. This can be obtained by adjusting the regulating switch to give 500 amperes at 160 volts. To leave the switch at this adjustment with the furnace empty would run its temperature up too high, so that if it were required to maintain a temperature of 1650 degrees F., while the furnace were standing idle for a short time, a readjustment of the switch would be made to give 200 amperes at 100 volts to make up the wall loss of 20 kilowatts when the doors are closed. If the doors are left open, enough more power would be required to make up the 8 kilowatts loss through that opening.

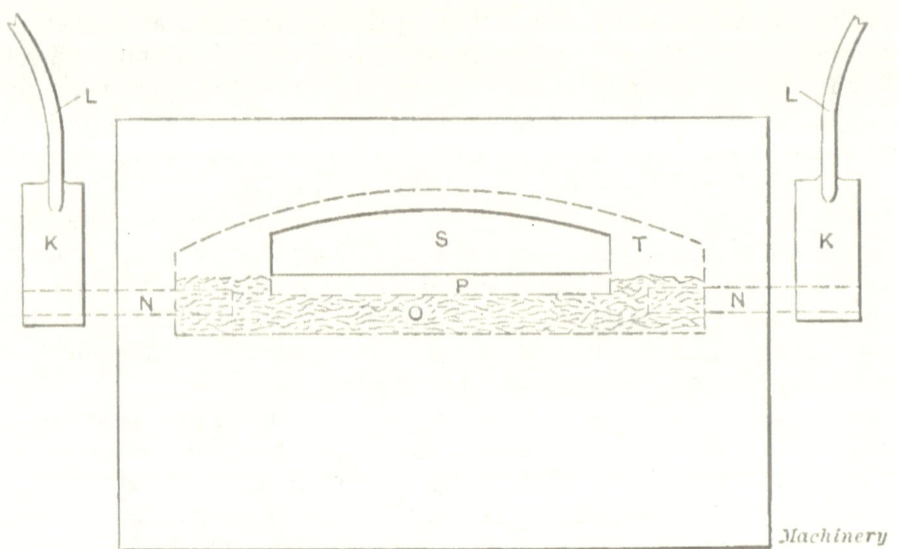

Fig. 1. Diagram showing the Principle on which the Baily Furnace operates

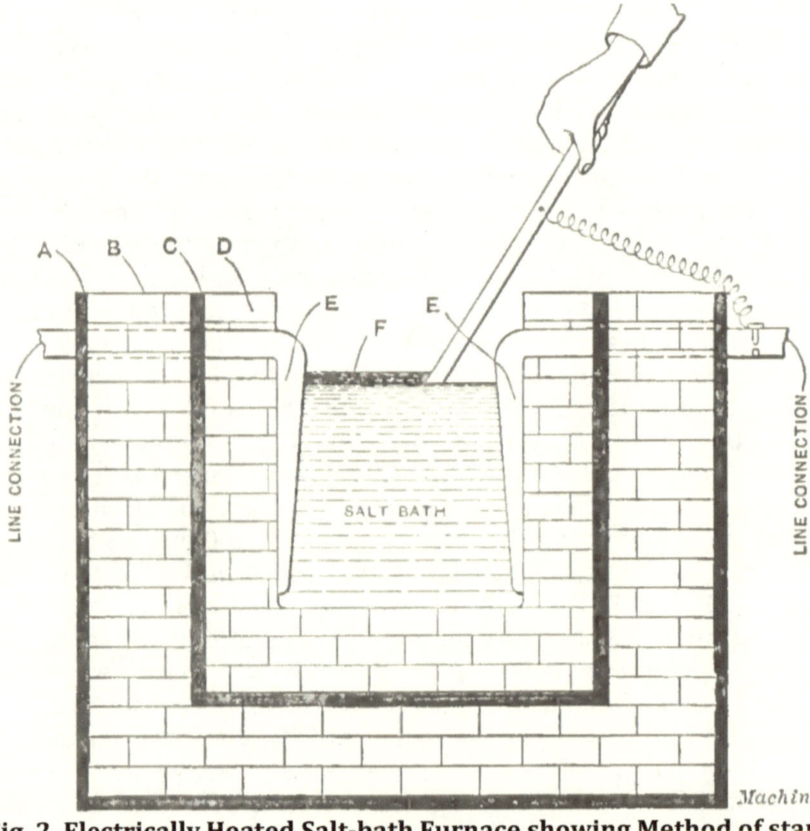

Fig. 2. Electrically Heated Salt-bath Furnace showing Method of starting with a Hand Electrode

A 60-kilowatt furnace of the same type was built for a forge shop in Alliance, Ohio. This furnace is operated every day to heat chrome-vanadium steel to a forging temperature of over 2000 degrees F., and allowed to get cold at night. It has a capacity of 300 pounds of steel per hour and is of the same type as the heat-treating furnaces. In fact, it could be operated just as efficiently at the hardening temperatures of the carbon or alloy steels.

The principle on which these furnaces operate is shown by the diagram, Fig. 1. Here *T* represents the heating chamber and *S* the opening or door leading into it, while *P* is the floor of the heating chamber on which the work is placed that is to be heat-treated. The leads which conduct the current to and from the furnace are shown at *L* and the copper plates that conduct it to the electrodes *N* are shown at *K*. A channel *O* running underneath the floor of the heating chamber is filled with a resistance material, which in this case is coke ground to about the size of peas. The electrodes *N* extend several inches into the ground coke, and as the electricity passes through the resistance material, from one electrode to the other, the required heat is generated. As many channels may be used as the size of the heating chamber makes necessary.

The Salt-bath Furnace. — The salt-bath furnace illustrated in Fig. 2 is of the type already described in Chapter Seven. The steel is immersed in a molten metal salt to bring it up to the proper hardening or drawing temperature. The noteworthy feature of this furnace is that it is electrically heated, and such a furnace can be made to last indefinitely. It is best to construct a sheet-steel shell *A* that is held together with angle irons, lined with one inch of asbestos and then built up with about 12 inches of common brick *B* on the bottom and four sides. This brickwork should again be lined with one inch of asbestos *C*, inside of which about 8 inches of firebrick *D* is laid on the bottom and four sides. When completed, the pot so formed should be of the right size for ordinary classes of work. It ought to be enough larger than the steel to be heated to allow space for two electrodes *E* of boiler plate on opposite sides of the pot, as these are placed inside of the salt bath, and also to allow for keeping the steel at least one inch away from the bottom, sides and electrodes, and immersed two inches below the top of the bath. When such a furnace has cooled down, the hard salt is not a good conductor of electricity. Therefore in starting up, it is necessary to chip a channel across the top of the salt, lay a ½-inch round carbon *F* in the channel, cover it with the salt chips and start melting them with a hand electrode, as shown in the illustration. When the channel is filled with molten salt, from one boiler plate electrode to the other, it will start the electric current flowing and the balance of the salt in the bath will soon become molten. The temperature is raised and controlled in practically the same manner as with the oven furnace previously described. The electrodes will last something like 3000 hours at temperatures that are high enough for hardening carbon steels.

As mentioned in the preceding chapter, when steel is heated to the hardening temperatures in a salt bath and then removed to be quenched, a thin coating of the salt adheres to the steel and prevents oxygen from attacking it and forming an oxide on its surface or raising a scale, while it is passing through the air to the quenching bath. This adhering salt cracks off when it is suddenly chilled in the quenching bath and leaves the steel with a natural color instead of with the black appearance that is produced when hardening it in other ways. At the higher hardening temperatures of high-speed steels, the salt bath furnaces that use graphite crucibles cause the tools heated in them to become pitted. In this electric furnace, however, the pot that holds the salt bath can be built up of other refractory materials, such as silicon-carbide and electrically calcined magnesia, and then no pitting occurs at any of the temperatures used for hardening.

Melting Points of Different Salts Used For Heat-treating Steel

Name of Salt	Melting Temp., Deg. F.	Name of Salt	Melting Temp., Deg. F.
Barium chloride	1580	Potassium carbonate	1526
Sodium chloride	1418	Sodium carbonate	1317
Potassium chloride	1346	Lithium carbonate	1283
Calcium chloride	1328	Potassium nitrate	644
Magnesium chloride	1306	Sodium nitrate	572
Lithium chloride	1112	Sodium silicate	113
Lead chloride	932	Barium fluoride	1832
Cupric chloride	928	Calcium fluoride	1832
Silver chloride	844	Magnesium fluoride	1664
Cuprous chloride	813	Sodium fluoride	1656
Ferric chloride	572	Lithium fluoride	1474
Zinc chloride	504	Potassium fluoride	1454
Aluminum chloride	356	Strontium fluoride	1350

Many different kinds of salts have been experimented with and the best kind to use for the bath depends on the temperature that is required. The melting points of the various salts that have been used are given in the accompanying table. Of the salts specified some are too expensive for commercial work, others volatilize too easily and still others cannot be used for various reasons. Several combinations can be made that are better than when one salt is used alone, and some of these combinations have a lower melting point than either of the salts forming the mixture. For temperatures between 1800 and 2400 degrees F., chemically pure barium chloride is without doubt the best salt to use, as it volatilizes less than any of the others and is low in price. For temperatures between 1400 and 1650 degrees F. three parts of barium chloride to two parts of potassium chloride give excellent results. For any of the drawing temperatures below 1075 and above 480 degrees F. equal parts of potassium nitrate and sodium nitrate make a salt bath that is satisfactory in every way. This can be kept molten at 400 degrees F. if it is continually stirred, but if left standing it will solidify at about 475 degrees F.

When the proper salts are used there is very little loss from volatilization and aside from the cost of current, the expense of operating this furnace is very slight. It is particularly adapted for treating small work that can be loaded into metal baskets or racks and then immersed in the molten salt, as large quantities can be handled in this way and the work is clean when finished. In one case, 70 pounds of safety razor blades are heated to the hardening temperature in this way. They are next quenched and then reheated to the drawing temperature in another electrically heated salt-bath furnace. Both furnaces are maintained at the correct temperatures for the hardening and tempering operations by a switch with eleven points that correspond to ten steps either way and thus gives twenty-one adjustments. A pyrometer tells when the bath is at the correct temperature, and then it is only a question of leaving the work in the bath long enough to reach a uniform temperature. The thickness of the work determines the length of time it should be kept immersed, and with a little experience this can be definitely determined.

Electric Arc Heating. — The electric arc has been successfully used for heating steel to hardening temperatures, and it is especially applicable for localizing the hardening in a certain part of the piece, the point of a cutting tool being a notable example. In Fig. 3 is shown a home-made apparatus that was rigged up for this purpose. The barrels filled with salt water take the place of a more expensive transformer and rheostat. By raising and lowering the steel terminal plates *A* in the barrels, the electric current can be controlled so that the tool is heated to the desired temperature. The carbon holder *B* is a very simple thing to make and any cast-iron plate *C* can be set on a rubber mat *D* to hold the tool *E* to be heated. A welding heat can also be obtained with this apparatus. It is essential that all parts of the body be protected from burns, as the heat from the arc will burn any exposed part in the same way that a sun glass burns on a hot day. Therefore it is necessary to wear gauntlets and a face shield with colored eye-glasses.

The steel is only heated in a spot directly under the carbon, and to heat the desired surface it is necessary to keep the carbon moving in a circle. It should not be brought too near the cutting edge of the tool and the arc should be started at a very low voltage, which is steadily increased to the desired point by adjusting the shunt rheostat. The carbon must also be kept a short distance away from the steel, for if it touches it is very likely to melt the metal at that point. The correct hardening temperature must be judged with the eye, as it would be very difficult to measure it with any kind of an instrument.

As steel becomes non-magnetic when it reaches the correct hardening temperature, or the transformation point, a magnet might be used to ascertain when it had reached the non-magnetic stage; but as steel cannot be instantly cooled from the hardening to atmospheric temperature, it must be

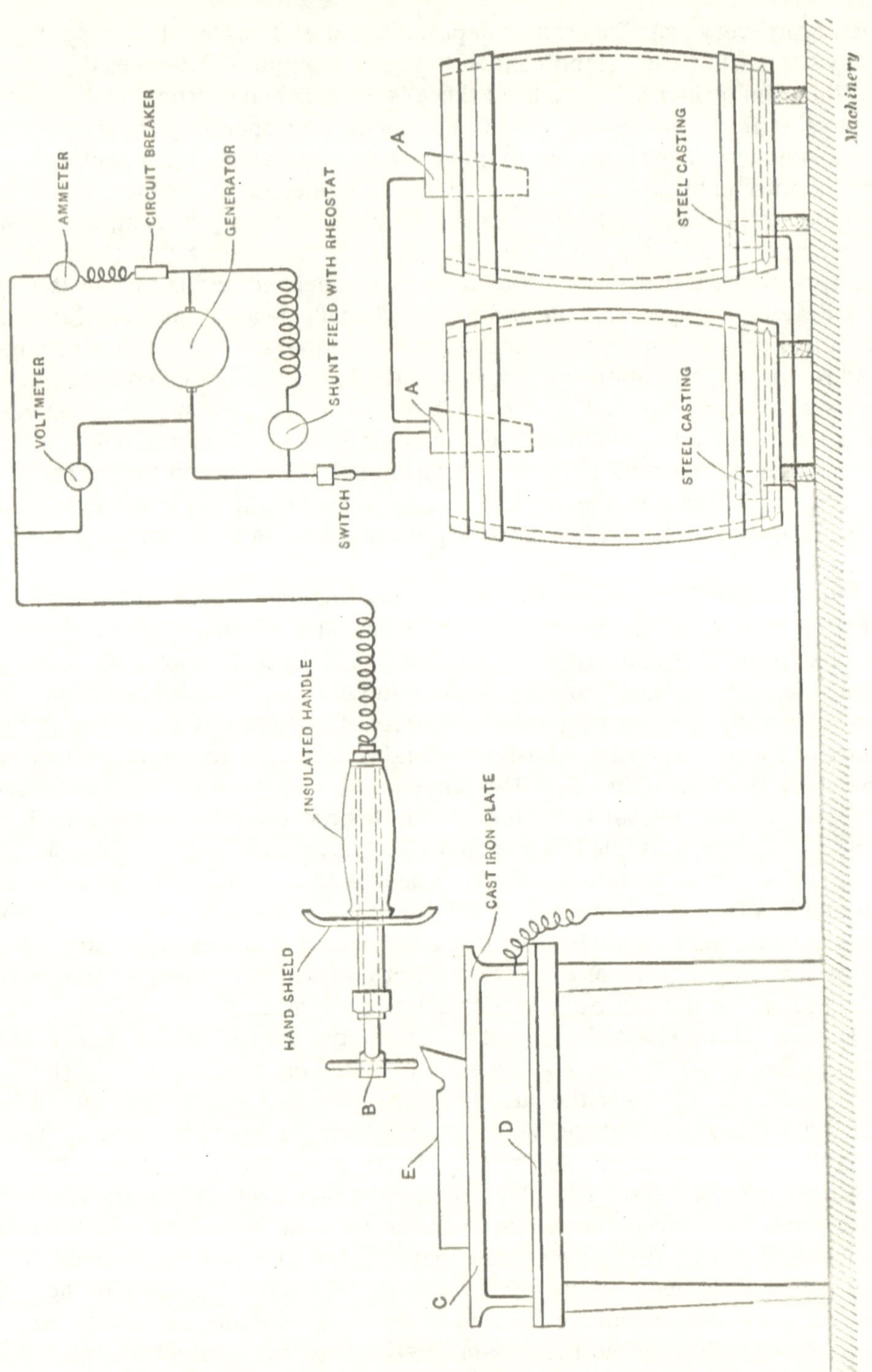

Fig. 3. Electric Arc Arrangement used for Local Hardening

heated to from 25 to 40 degrees above the non-magnetic point to allow for the lag when quenching. It is very difficult to judge these 25 to 40 degrees with a magnet, and it cannot be used for accurate work in hardening.

The rapidity with which a piece of steel can be heated to the hardening temperature is the greatest recommendation of the apparatus just described, as it takes only two or three minutes to heat quite a large surface on a fairly thick piece. If it were not heated and quenched quickly, the oxygen in the air would have time to raise quite an oxide or scale on the heated steel. If this principle were used regularly, it would be much better to fit the apparatus with a transformer and switchboard control in place of the water barrels, as these require constant attention to keep the water from boiling over, and to raise and lower the terminal plates.

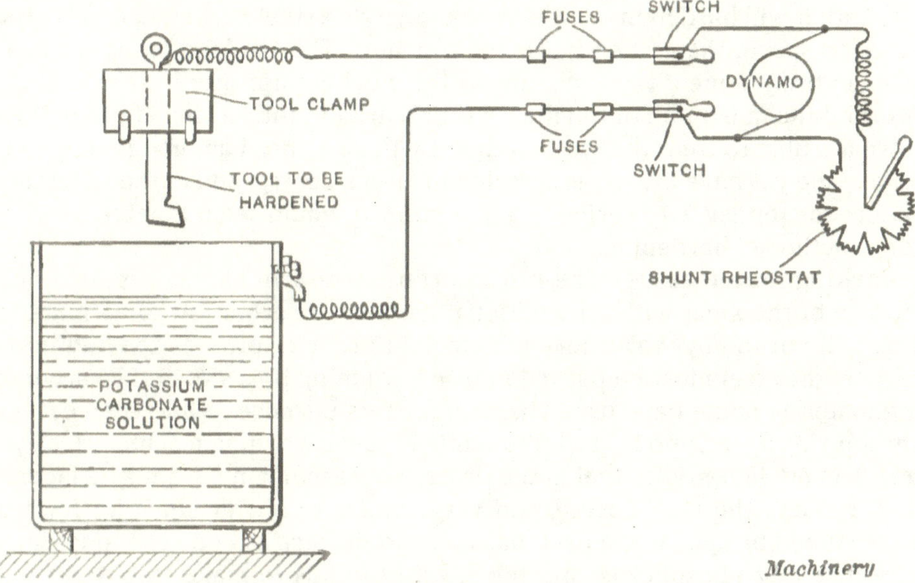

Fig. 4. A Rapid Method of Hardening Steel

The most rapid method of hardening steel is doubtless that in which the furnace shown in Fig. 4 is used. This consists of a cast-iron tank containing a potassium carbonate solution, a clamp in which to hold the piece to be hardened, and a rheostat, switches, fuses and wiring. After clamping the tool, it is only necessary to turn on the current, lower the tool into the solution and, when it has attained the proper temperature, turn off the current and allow the steel piece to be quenched by the solution. When the steel enters the potassium carbonate solution, it completes the circuit and immediately begins to heat up. The correct temperature is reached on a good-sized piece in about a minute and, being quenched before it is removed from the bath, nothing can attack the steel to discolor it. Consequently, when taken from the bath it has a clean steel-colored appearance. Being heated uniformly on all sides, there is no tendency for the work to warp or become distorted, and it can be lowered into the bath only as far as it is desired to have it hardened. With a little experience, the rheostat can be set at the point that will heat the steel to just the required temperature, and then it is only a question of leaving the steel in the bath long enough to attain this temperature before turning off the current. For permanent use, this apparatus can be fitted up to enable the current to be quickly regulated for obtaining any desired temperature.

Chapter Nine - Miscellaneous Hardening Methods

Pack-hardening. — Pack-hardening, as the term is generally understood, consists in treating steel, generally tool steel, with some carbonaceous material until it will harden in oil. It is well known that steel hardened in oil is less liable to spring than when hardened in water. The tendency to crack is almost entirely done away with, unless the steel is improperly treated in the fire, and the maximum of toughness is procured in the hardened parts. Now, if we are able to treat the steel so that it will be as hard as though dipped in water, and yet have the toughness due to oil-hardening, and at the same time reduce the tendency to spring to a minimum, it would seem that we have the ideal method of hardening.

Packing Materials. — The process consists essentially in supplying the surface of the steel with an additional amount of carbon by some material that will not in any way injure the steel. In order to provide the additional carbon, the steel must be packed in iron hardening boxes with the carbonizing material. Some have used charcoal for this purpose. While charcoal is a carbonizing agent, and is used frequently in casehardening machine steel, yet its effect on high-grade steel in the process of carbonizing is not satisfactory, as it renders the steel coarse, and very similar to blister steel. No form of bone should be used when pack-hardening tool steel, as bone contains a high percentage of phosphorus, and the effect of this is to make steel weak and brittle.

For steel that does not contain more than 1.25 per cent carbon (125 points), charred leather gives the best results. Above this percentage use charred hoofs, or horns, or a mixture of the two. The leather, hoofs, or horns, may be used over and over by adding a quantity of new material each time.

Method of Packing. — The work should be packed in the hardening boxes so that no part of any piece of work comes in contact with the boxes; in fact, there should be at least ½ inch space between the work and the box. A layer of the carbonizing material should be placed in the bottom of the box, and a layer of work placed on this, taking care that no two pieces touch each other. If we are treating gages, or pieces of steel that are apt to spring unless care is used, we should make sure that they are so placed in the box that there will be as little liability of springing as possible when they are drawn up through the packing material. They must not be dumped into the hardening bath, as is the case when ordinary casehardening is done.

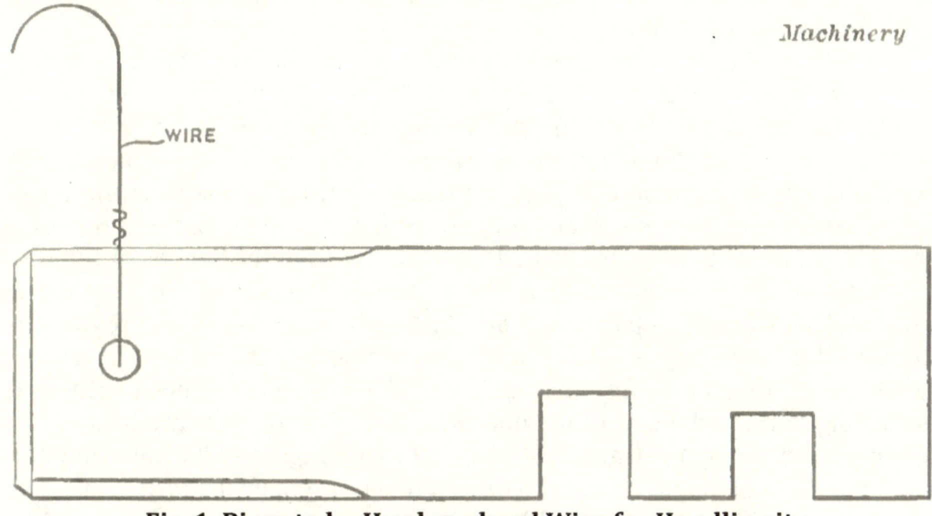

Fig. 1. Piece to be Hardened and Wire for Handling it

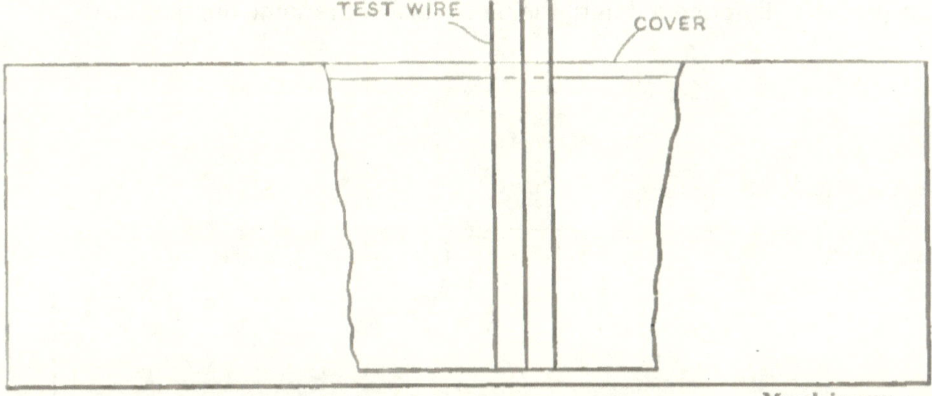

Fig. 2. Diagrammatical View of Hardening Box with Test Wires which enable the Operator to determine the Temperature within the Box

In order to be able to properly handle the work, each piece should be wired with a piece of *iron* binding wire, as shown in Fig. 1, and the pieces so placed in the box that there will be the least resistance possible when drawing them out. At times they may stand on edge, as shown in Fig. 1. For certain shapes, however, it is advisable to stand them on end.

When several layers of work are packed in a box, the wires should be so arranged around the edge of the box that the various layers may be taken out in order, commencing with the top row. This is easily accomplished by marking the sides of the box with chalk, designating the side where the top row of wires is, as 1, the one where the second row is, as 2, and so on. Unless we adopt some such method, the pieces get all mixed up, and some will be drawn to the surface of the packing material, and will cool before the operator has a chance to dip them.

Heating the Steel. — As in all heat-treatment of tool steel, the heat should be as low as is consistent with desired results, and the heat must be uniform throughout the box. It is also necessary that we gage the length of time the steel is exposed to the action of the carbonaceous material. Unsatisfactory results follow any attempt to gage the length of time by the time the boxes are in the furnace. In order that the operator may know when the contents of the box are heated, holes are drilled through the cover of the box at the center, and test wires are run down to the bottom of the box, as shown in Fig. 2. These wires should project about one inch above the top of the cover. The holes in the cover may be of any size to accommodate the wire to be used; a good size is ¼-inch hole for 3/16-inch wire. When the box has been in the fire, according to the judgment of the operator, until the contents are heated to a low red, a wire may be drawn, by means of long tongs, and its condition noted; if it is red hot, begin timing the heat; if it is not red, wait a little while, and draw another. Continue doing this until one is drawn that is of the desired temperature. The wires passing down at the center of the box, and between the pieces, will not be red until the pieces are of the same temperature.

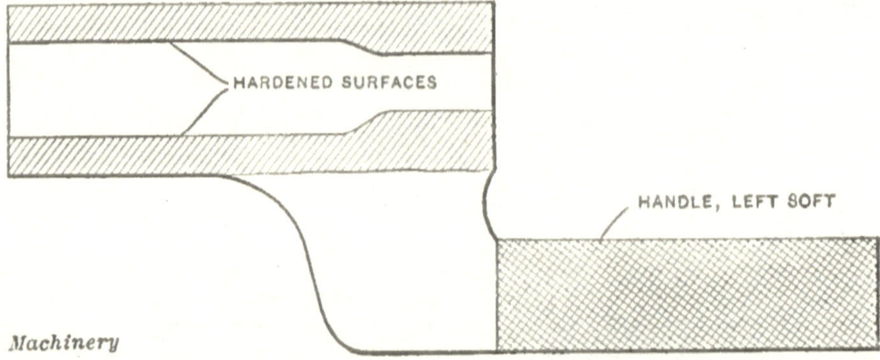

Fig. 3. A Gage Hardened on the Surfaces Indicated

The length of time necessary to expose the pieces to the action of the heat depends upon how deep we wish to harden the steel. For ordinary snap gages 1½ to 2 hours after the steel is red hot is sufficient, but the time must be varied according to the percentage of carbon that the steel contains and its intended use.

Pack-hardening Gages. — Sometimes locating gages are made with the gaging holes made to the finished size of the gage. This method is not to be advocated where it is possible to use hardened bushings. In the latter case, the holes in the gage may be made of the proper size for the bushings, and the gage left soft, while the bushings are hardened, ground and lapped, and pressed into place without any tendency to distort the gage. But when it is necessary to make the gage of one piece, and have the gaging holes to size in the gage without bushings, then the pack-hardening method will be found to work satisfactorily, as the heat may be very low, and the tendency to distortion will be eliminated, provided the processes of annealing and machining have been properly done.

As an example of pack-hardening gages, a case from practice may be cited. The gage was of the form shown in Fig. 3, and it was necessary that the walls of the opening through the gage be hard, yet the opening must retain its shape. The hole was filled with finely-pulverized charred leather; the handle and the portion connecting it with the body were covered with fireclay which was wound with fine iron binding wire to prevent it falling away when baked. The gages were packed in scale collected in the forge shop. This scale came from the outside of pieces of iron and steel as they were being forged. Being free from carbon, they absorbed or took the carbon from the surface of the steel. The ends of the opening through the gages were covered with fireclay mixed with water to the consistency of dough, which was allowed to harden before the gages were packed. The fireclay prevented the carbon gas escaping from the leather, as the scale would have taken it up very quickly.

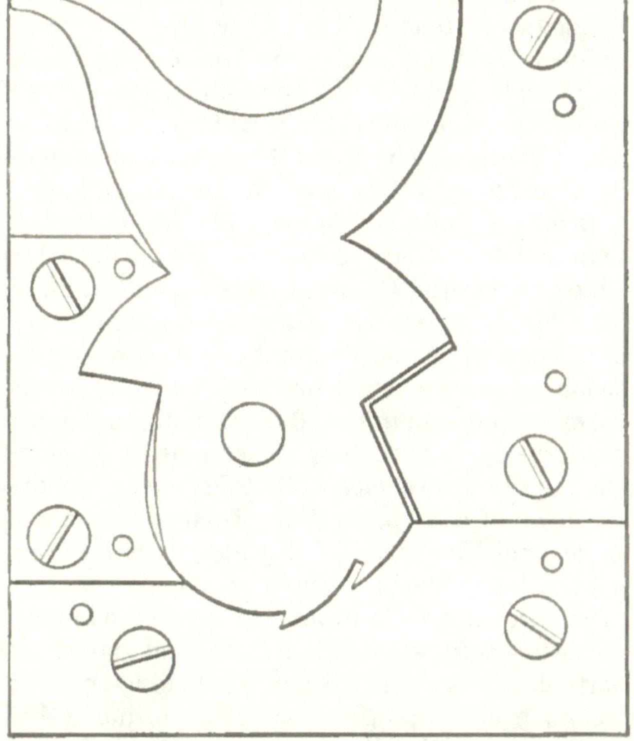

Fig. 4. Receiving Gage of a Type which is Hardened to Advantage by the Pack-hardening Method

When the gages had been exposed to the carbonizing influence of the leather for one and one-half hour, they were removed from the box in which they were heated, placed in a bath of raw linseed oil, one at a time, and a jet of oil was pumped through the opening after the leather had been removed. The fireclay around the handle and the portion connecting it with the body was left on until the gage was hardened, when it was removed. The walls of the hole were found to be hard, and as the surface of the gage was practically decarbonized, there was little danger of its pulling the piece of steel out of shape. The handle and adjoining portion, being protected by the fireclay, did not harden, or even cool quickly enough to distort the gaging portion in any way.

Often receiving gages are made of several pieces which are fastened to a plate as shown in Fig. 4, which is a receiving gage for a gun hammer, and it is necessary that the various portions be gaged accurately, and that each portion bear a certain relation to every other portion.

As shown, the various portions of the gage are made in sections, fitted in place and hardened. Unless these pieces are hardened by some method that eliminates the tendency to spring, they will be of little use after they are hardened. This is a case for pack-hardening. Pack the pieces in leather in a small iron box, run for one hour after they reach a low, red heat, and harden in raw linseed or sperm oil. It will not be found necessary to heat the steel treated in this way as hot as if heated in an open fire and dipped in water. It is not necessary to heat steel in the form of gages quite as hot as if it were made into cutting tools; but, even in the latter case, be sure to keep the heat down, and do not dip in extremely cold oil; have it warm, but not hot.

Pack-hardening Hammer Dies. — The hardening of large die-blocks, for either hot or cold work, is a serious problem to most manufacturing concerns, because, in addition to the cost of the steel, a very heavy expense for the labor of die-cutting is involved. The total cost of a pair of dies frequently ranges from one hundred to three hundred dollars or more, so that the loss of a single die in hardening is a serious item of expense, aside from the long time required to replace it. Furthermore, a hammer die that is subjected to the blows of a drop, weighing from twelve hundred to eighteen hundred pounds, must be very skillfully heat-treated in order to stand up under the repeated heavy shocks without breaking.

The requisites for a properly hardened hammer die are: a perfectly hard face, with a sufficient depth of hardening to prevent the surface from sinking; sharp, clean edges and corners; a soft, tough body to insure against breakage; freedom from warping or change of size, due to shrinkage of the steel in cooling. The face of the die should not only file hard, but should stand a smart blow of a hardened steel hammer without perceptible dent. With any good uniform die steel these qualities may be readily obtained by the following treatment.

Furnaces for Heating. — For heating the dies, an oil or gas furnace is the ideal equipment, but a coal or coke-fired muffle, capable of maintaining a temperature of at least 1600 degrees F., will answer the purpose, provided the temperature can be held constant. The great advantage of the gas or oil furnace over the coal or coke-fired muffle is the greater rapidity with which the temperature can be raised or lowered, and the ease with which it can be held constant.

A temperature-indicating device, either pyrometer or pyroscope, is an absolute necessity if good results are to be obtained, as it is impossible to judge the temperature by the eye with sufficient accuracy for hardening this class of dies. A temperature difference of twenty-five degrees has a very noticeable effect upon the die, while a fifty-degree variation may spoil it entirely, particularly if at the critical high heat demanded with some of the best alloy steels. Under certain conditions, a temperature difference of even seventy-five degrees cannot be detected by the eye with certainty. No guesswork should be permitted to enter into the hardening of dies, as a few failures cost many times the price of a good pyrometer or pyroscope.

Packing the Dies. — Dies should always be packed for hardening. To pack a die properly, mix a thick paste of linseed or cottonseed oil and powdered bone-black. Paint the face of the die with a thick coat of this paste, which will bake on under the action of the heat and protect the delicate edges from oxidation through contact with the air.

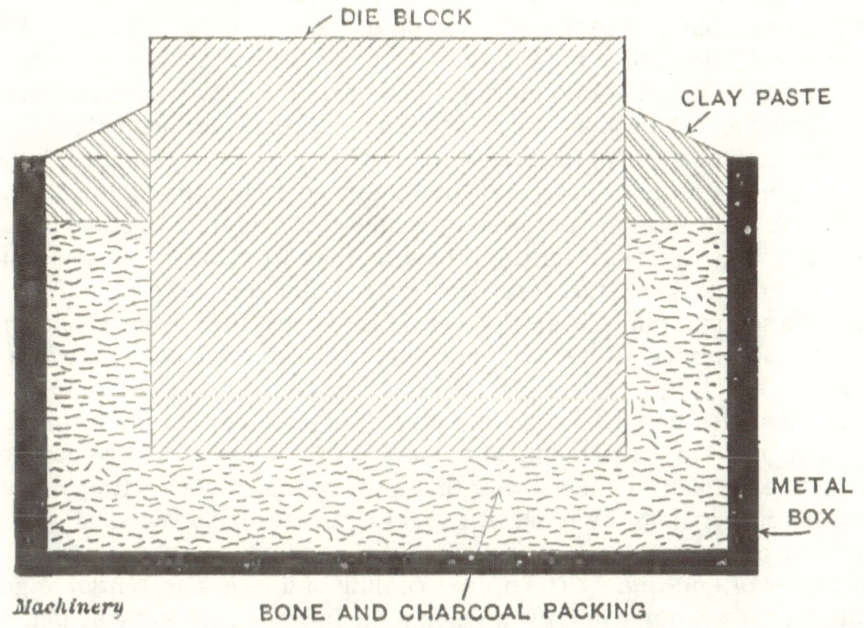

Fig. 5. Packing a Die-block for Heating

Next, take a shallow sheet-iron or cast-iron box, an inch or so wider than the die all around, and fill the bottom to the depth of an inch with fresh bone and powdered charcoal, mixed half-and-half. Place the die, face down, in the center of the box, taking care not to displace the paste on the face. Now fill in between the sides of the box and the die with more bone and charcoal, mixed half-and-half, right up to the top of the box. The die should set in the box only deep enough to allow its top edge or back to project about an inch above the top of the box, for convenience in handling with the tongs when removing from the box to quench. Now cover the space between the upper edge of the box and the sides of the die with a thick layer of wet clay paste, which will bake on and keep the charcoal from burning out, and the die is ready for the furnace. Fig. 5 shows the arrangement of the die and packing in the box.

Method of Heating. — Die steels in general give the best results when quenched at the temperature which produces the densest, finest-grained fracture. Unless this temperature is exactly known for the steel used, first ascertain it by placing a number of small pieces of the steel in the furnace (they do not need to be packed for this purpose) and hardening and fracturing them at different heats, starting at a low heat and gradually working up until the greatest refinement of the grain of the steel is obtained. Note the temperature at which it is obtained, as that will be the correct quenching heat.

The dies should be put into the furnace as soon as it is lighted and allowed to "come up" with it. If the correct quenching temperature for the steel is, say, 1500 degrees F., then the furnace should be checked when the pyrometer indicates 1400 degrees, and the dies allowed to soak at that even heat for three or four hours. Then the heat should be slowly raised to the final 1500 degrees, and held at that point for one or two hours longer, according to the size of the die, large blocks requiring a longer time to attain an even temperature throughout than small ones. Then the die is ready for quenching.

Do not attempt to hurry the heating of dies; as it takes several hours for the heat to penetrate the packing and thoroughly soak into the die-block, until the outside and center are of the same temperature. More dies crack in cooling, or afterwards, from insufficient heating than from any other one cause. With this in mind, five hours is the least permissible total time in the furnace, and seven to eight hours is much safer. Most expert die hardeners put their dies into the furnace in the early morning, and take them out in the late afternoon.

Care must also be taken that the dies get an even heat in the furnace. As may readily be proved by manipulation of the pyrometer rod, some parts of a furnace are frequently much hotter than others; so, to avoid getting one end of the die hotter than the other, it should be occasionally reversed, end for end, while heating.

Methods of Cooling. — The proper cooling of dies is as essential to good results as the heating itself. Even when a die has been correctly heated, improper cooling may make it too soft, or warp it, or even cause it to crack, rendering it totally unfit for use. The requisites for properly cooling a die are: a sufficient flow of cold water to cause it to harden to the necessary degree of surface hardness to withstand wear, and to a sufficient depth to prevent the surface from sinking under heavy blows of the hammer; and means for applying the water in such a manner as to cause the least possible warping. The rate of cooling has an appreciable effect upon the initial heat required to make a die hard enough. If the water equipment permits of very quick cooling, a somewhat lower heat may be used, and the lower the heat, the less trouble from warping. It is useless to attempt to cool a die-block of any appreciable size in a tank of still water, as the heat in the large body of metal draws the surface faster than the water can harden it. It will harden upon the immediate surface, under such treatment, if heated enough, but there will be no depth to the hardening, and the face will soon sink in use, causing cracks to form. A properly hardened die should have a depth of hardening of from ¼ to ¾ of an inch, according to the steel used. Oil should not be used to harden hammer dies, as its cooling action is not sharp enough to produce a sufficient depth of hardening.

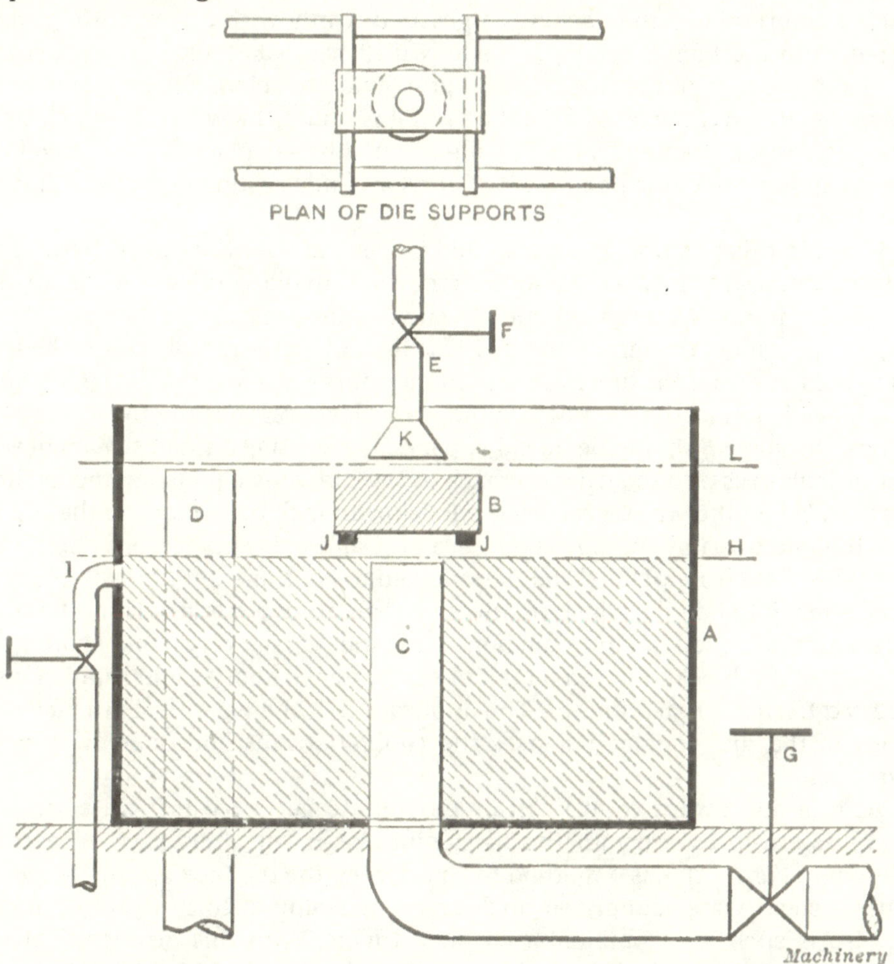

Fig. 6. Cooling Bath Arranged to throw Water against both Faces of the Die

In cooling, the face of the die should receive a sufficient flow of cold water to cause it to harden to the greatest possible depth. The back of the die should at the same time receive a similar treatment, to make the shrinkage of face and back equal, thereby preventing warping. The sides of the die may be left to take care of themselves. Dies may be hardened either face up or face down, according to the equipment used.

Cooling Baths. — Fig. 6 shows a most desirable arrangement for hardening dies face down. *A* is the cooling tank; *B*, the die; *C*, a six-inch water pipe; *D*, an eight-inch overflow pipe; *E*, an inch-and-a-half water pipe; *F*, a quick-opening valve; *K*, a flattened nozzle with a long, narrow opening; *G*, a quick-opening valve; and *J*, adjustable rods for supporting the die at the two ends.

To harden a die, the water level is lowered to the line *H*, by means of the two-inch drain *I*. The packing box containing the die is then removed from the furnace and set upon the floor. The back of the die, which protrudes an inch or more above the packing, is grasped firmly with a pair of tongs and lifted from the box. If the bone paste, which has baked on in a hard crust, sticks to the face of the die, no attempt is made to remove it, as it serves to keep the air from the face of the die until it is under the water, at which time the paste will crack and fly off. The die is immediately placed in the tank, face down, under the elongated nozzle *K*, the two ends resting upon the adjustable supports *J*.

When in this position, the die should be central over the pipe *C*, as well as central under the nozzle *K*. As soon as the die is in place, the valve *G* is turned on, which permits a large volume of water under pressure to flow out of the pipe *C* up against the face of the die. This flow of water rapidly raises the level in the tank from the line *H*, so that in the course of a few seconds the whole die-block is immersed. Just before the water level reaches the top of the die-block, the valve *F* should be quickly opened, permitting a small stream of water at high pressure to strike the back of the die, thus equalizing the cooling strains. When the water reaches the level *L*, an inch or two above the top of the die, the overflow pipe *D* takes care of it and prevents it from rising further. The flow from pipes *C* and *E* is allowed to continue until the die-block is cool enough to handle, which requires only about two minutes from the time the water is turned on, for a medium-sized die. This is very rapid cooling, as this same die-block, if put in a tank of still water to cool, would retain a visible red for five or six minutes. The application of the water to both face and back of the die, equally, balances the cooling strains and does away with warping.

As it is not always convenient or possible to get a six-inch water supply even at low pressure into the hardening room, an alternative scheme is shown in Fig. 7. This is a method for hardening the die face up, and requires only a small water supply, an inch-and-a-half pipe at forty to fifty pounds pressure, such as is obtainable in most towns. With this arrangement we have the tank *A*, the iron plate *B*, supporting the die-block *C*, the inlet pipe *D*, fitted with the nozzle *E*, and the overflow pipe *F*.

In operation, the water is lowered by means of the drain *J* to the level *I*, which covers the plate *B* about an inch deep. The die is removed from the

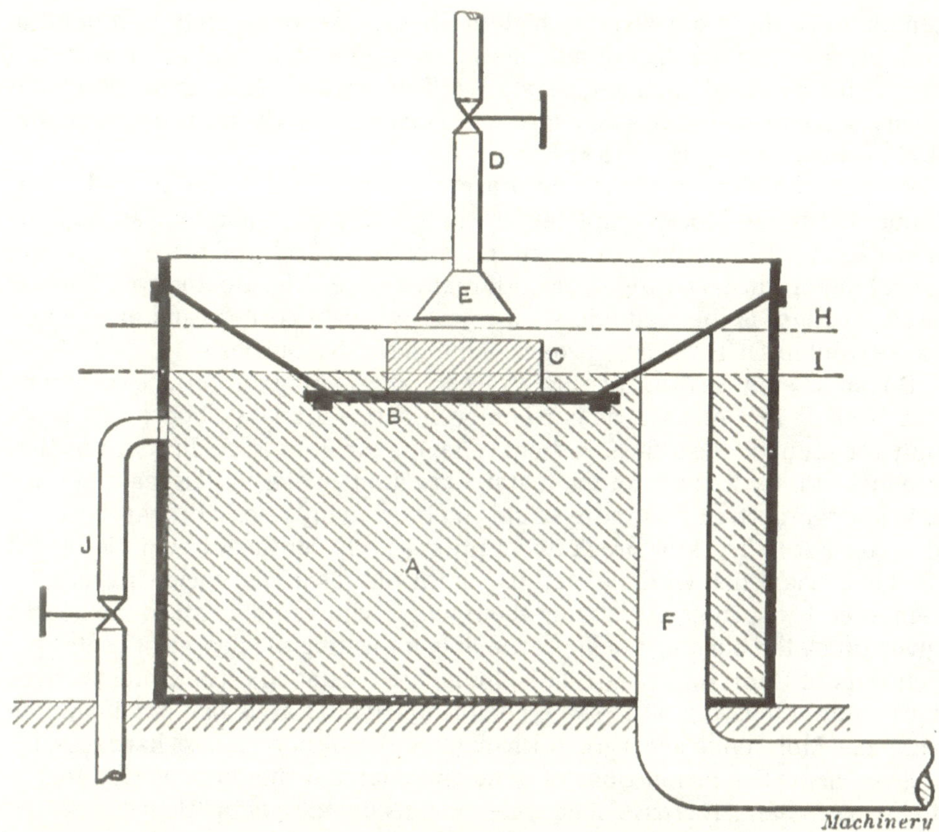

Fig. 7. Cooling Bath Arranged to partially Immerse the Die and to throw Water against the Top Face of the Die

furnace as before, and placed upon the plate *B*, centered under the nozzle *E*. The water that covers the plate, and the cold iron itself, rapidly cools the back of the die. As soon as the die is in place, the valve *D* is opened, and the water from the nozzle *E* strikes the face of the die. The water level in the tank rises rapidly to the level *H*, completely immersing the die, and the overflow pipe *F* holds it there. The water from the pipe *D* is permitted to flow upon the face of the die until cool enough to handle.

With this last arrangement, the flow of the water from the pipe *D* must be so regulated as not to cool the face faster than the back of the die; otherwise warping will occur. With a little practice this method will produce nearly as good results as the double-pipe arrangement, except that the dies will have a little greater tendency to warp. Either method, though, is far ahead of the common tank in this respect.

Tempering in Oil. — Dies should be tempered, or drawn, as soon as they are cool enough to remove from the tank. If not drawn at once, they are liable to develop cracks. For tempering, the proper equipment consists of a small tank of oil, a suitable burner underneath for heating it, and a high-temperature thermometer. Any high-grade cylinder oil of high flash point is suitable for the work. Low-grade oils smoke unpleasantly at moderate heats, and will not stand high temperatures. The drawing temperatures of die steels range from 350 to 480 degrees F. The color barely starts at 350, while 480 produces a very full straw.

Heat the oil bath to the temperature corresponding to the desired color, immerse the die-block completely in the oil, and let it soak for an hour or more. Remove from the oil, quench in water or cool oil, and place in a warm, even temperature over night, or until ready to use. A lead bath should not be used for dies, as the lead will stick to the intricate corners and angles and cause trouble. Oil is the only proper medium for this purpose.

Do not attempt to draw die-blocks by simply starting the surface color on a hot plate or in a furnace, as is common practice in many shops. This heats only the immediate surface, without relieving the greater cooling strains that are locked into the body of the metal. For hammer dies under heavy service it is most important that these strains be thoroughly released; otherwise the dies are apt to crack or break down prematurely. Surface-drawn die-blocks should be classified with the methods of the old-fashioned spring maker who tempered his springs by rubbing them with a charred pine stick, the sparks given off by the burning wood presumably indicating the degree of heat.

Points to Remember. — 1. Fresh bone has a decided carbonizing (surface hardening) effect upon the steel. The ordinary mixture is half bone and half charcoal. More bone gives greater hardness, more charcoal less hardness, for a given heat. The proportions of bone and charcoal should be varied to suit the work in hand. Increase of hardness means increase of brittleness.

2. Unless the face of the die is treated with a jet of water under pressure, the sunken parts of the pattern will not harden equally with the face. When dipped into still water to harden, steam forms in the die cavity, which keeps the water from entering to properly harden these parts of the die. To overcome this, the water must be forced into the sunken parts of the die by pressure sufficient to overcome the resistance of the steam that is formed by the heat.

3. The most skillfully hardened dies will not give good service unless set absolutely parallel as to their surfaces under the hammer. In a block that is set in the hammer or anvil a little "off" the strains created are enormous, and breakage is sure to result speedily.

4. Hammer dies must have *depth* of hardening, as well as surface hardness, to withstand the heavy blows to which they are subjected. Other kinds of dies do not need this, because the work is light, and service is merely a matter of wear of their edges. Sinking of the face of a hammer die means the development of cracks every time, so the steels used should be selected for their deep-hardening qualities. This means alloy steels, generally, as few carbon steels will harden to a depth of more than 1/16 inch, which is entirely inadequate for this service.

To Prevent Scale on Dies when Hardening. — Scale can be prevented from forming on dies when hardening, if the dies are dipped in water before they are heated and then put into dry salt, letting all the salt that will cling to the dies remain on them. After this, the pieces are heated, and immersed in the quenching bath as usual. The scale or crust of salt will fall off in the bath and the piece so treated will have the appearance of a piece which has been heated in cyanide.

Heat-treating Dies and Tools used in Forging Machines. — Vanadium alloy steel has been found very suitable for dies for forging machines. Two grades of vanadium tool steel are recommended for forging machine dies by the American Vanadium Co., of Pittsburg, Pa. One contains 0.50 per cent carbon; from 0.80 to 1.10 per cent chromium; from 0.40 to 0.60 per cent manganese; and not less than 0.16 per cent vanadium. The silicon content should not exceed 0.20 per cent.

The heat-treatment recommended for this steel is as follows: Heat to 1550 degrees F. and quench in oil; then reheat from 1425 to 1450 degrees F. and quench in water, submerging the face of the die only. When this method is used, the die is drawn by the heat remaining in the body and is thus tempered and the life increased.

The second kind of vanadium steel recommended contains from 0.65 to 0.75 per cent carbon; from 0.40 to 0.60 per cent manganese; and not less than 0.16 per cent vanadium. The silicon content should not exceed 0.20 per cent. The heat-treatment for this steel should be as follows: Heat to 1525 degrees F. and quench in water with only the face of the die submerged. The length of life of vanadium steel dies is stated to be about six times the life of dies made from ordinary high-carbon tool steel.

When ordinary carbon tool steel is used for forging machine dies, these should be hardened as usual and the temper should be drawn so that they can be just touched with a file, or, in other words, to a light straw color.

Heat-treatment for Vanadium Tool Steel. — For a quenching bath, either water, oil, brine or sulphuric acid may be used, but oil is preferred if the best results are to be obtained. When using oil as a quenching bath, heat the steel to between 1450 and 1550 degrees F. (dull cherry-red) and then quench in oil. Draw the temper to about 425 degrees F. (very faint yellow). When using water as a hardening bath, heat to about 1400 degrees F. and quench.

Hardening the Heads of Forge Tools. — In the interests of safety and economy, it is advisable to harden the heads of all forge tools, such as sets, chisels, fullers, swages or any other tools made of high carbon steel, that are used by being struck upon their heads with a hammer. Upon the face of it, this may not appeal very strongly to a great many users of such tools, as it is common practice to leave the heads just as soft as possible. Some have even recommended annealing the heads from time to time to keep them soft, the idea being to prevent burrs breaking off and flying around when the heads get battered down from use. This is a mistake, for when soft steel is subjected to hammering for any length of time, it not only begins to batter down and burr over on the edges but also becomes crystallized. This makes the metal very brittle — although it is quite soft — with the result that the burrs begin to break off and fly around when the tool is in use. This shortens the life of the tool.

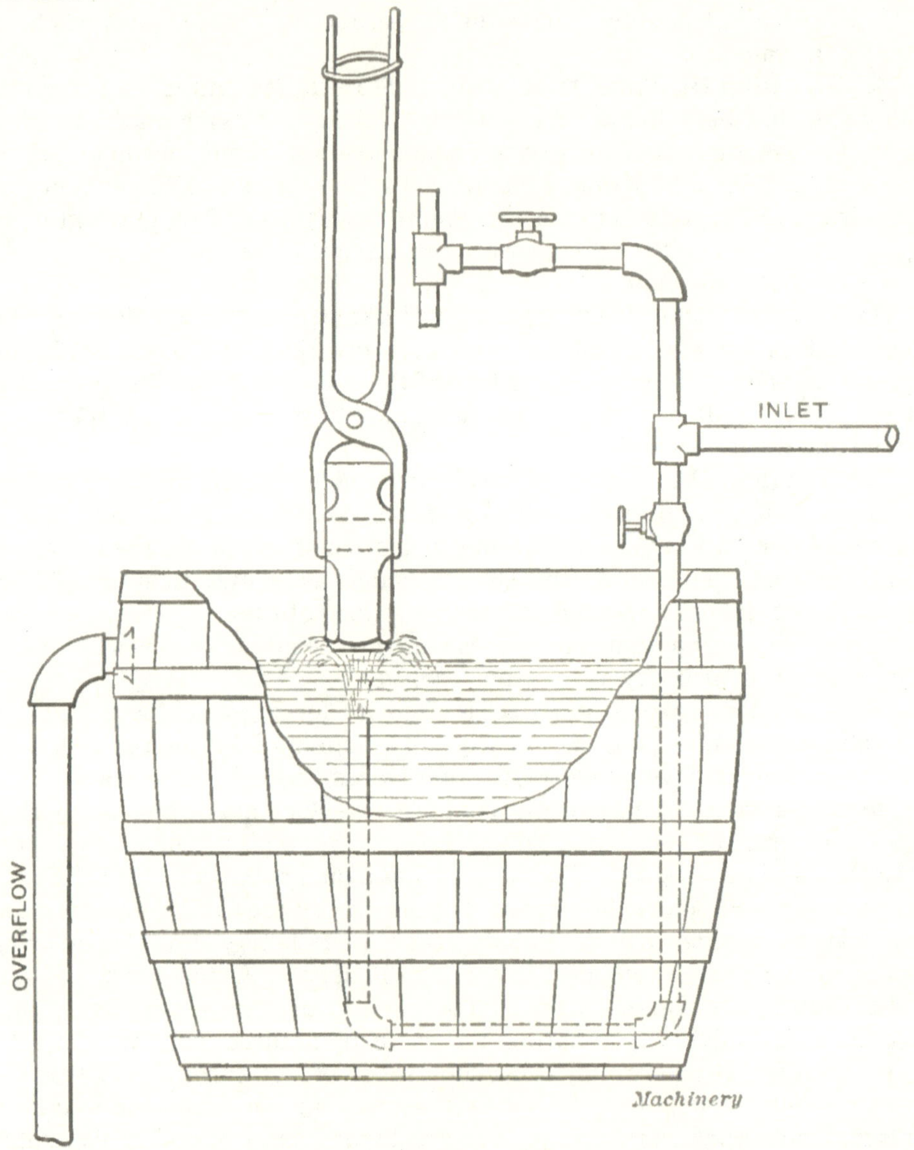

Fig. 8. Arrangement of Piping in Quenching Tank for Hardening Forge Tools

Experience has proved that high carbon steel can be hardened and tempered to make it suitable for any purpose from cutting tool steel in an annealed condition to withstanding the impact of blows. Therefore, theory, as well as the results of practical experience which will be given later, favors hardening the heads of tools. The best and safest type of head to use, either in a soft or hardened condition, is dome-shaped without a single sharp corner.

For a number of years, a user of these tools has made a practice of hardening forge tools on both ends with very good success. An idea of how these hardened tools will wear may be obtained from the fact that a few tools have been in constant use for upwards of eight years. A sledge nearly nine years old has been in use practically every working day since it was made and is still in perfect condition, not having even been reground on the face. A 17/8-inch swage, which is about eight years old, has been in constant use the whole time and is still without perceptible signs of wear. A rectangular set, which is one of the tools that a blacksmith uses the most, is about five years old, and like the sledge it is still in perfect condition.

The method of hardening a sledge or hammer of any kind is to first heat the peen to as near the point of decalescence as possible and then quench it in water in an ordinary slake tub, until the heat has been so far withdrawn that it will "carry water." The tool is then polished with a piece of wood covered with emery cloth and the temper drawn with the back heat until it shows a light copperish color, after which it is cooled off. The face is next heated in the same manner as the peen but it is cooled in a stream of water rising straight from the bottom of the quenching tub and striking the center of the face, as shown in Fig. 8. This insures the center being equally as hard as the edges, if not harder, as steam cannot generate and form a cushion as it may do where the tool is immersed in the water. When the face is fairly cooled to a depth of about 1 inch, it is polished and laid in a hot fire, which in a very short time draws the temper on the outer edges to a blue color, leaving the center just as hard as possible. When a hammer has been properly hardened in the manner described, there is practically no danger of its cracking or burring, as the hard center is supported on all sides by the softer and tougher metal around the edges.

The faces of fullers, swages, sets, etc., are hardened in exactly the same manner as the face of a hammer, the heads being heated slightly above the critical point and quenched to a depth of about ½ inch. They are then polished and the temper drawn by the back heat until the color has all but disappeared. What color is left may be described as light sea-green. Another gage for drawing the temper of the heads of tools is to let the temper run until the piece is hot enough to freely ignite thin scrapings of hickory wood; it is then dipped in water, just enough to prevent the temper from running any further, and allowed to cool in the air.

Heat-treatment of Spring Steel. — Experiments have been made on spring steel of the following composition: Carbon, 1.01 per cent; manganese, 0.38 per cent; phosphorus, 0.03 per cent; sulphur, 0.03 per cent; silicon, 0.13 per cent. These tests indicated conclusively that the elastic limit of one-percent carbon steel can be made to vary from 78,500 pounds per square inch to

240,000 pounds per square inch by changes in the heat-treatment, and that very small changes in the temperature for the drawing of the temper are sufficient to affect the elastic limit of the steel. The highest elastic limit is obtained by heating to a temperature of 1425 degrees F. and quenching in water, and then redrawing to 750 degrees F. Better all-round results, however, can be obtained without drawing the temper, by heating to a temperature of 1450 degrees F. and quenching in oil. In fact, when quenching in oil, a higher elastic limit is obtained by not drawing the temper at all. When quenching in water, the steel becomes too brittle for use as springs, unless it is drawn to a temper not less than 600 degrees F.

Heat-treatment of Screw Stock. — The composition of common screw stock is usually as follows: carbon, 0.08 to 0.25 per cent; manganese, 0.30 to 0.80 per cent; phosphorus, not over 0.16 per cent; sulphur, 0.05 to 0.15 per cent. It is an unsafe steel to use for any parts which require great strength or toughness. Screws made from it should always be heat-treated and should not be used in the annealed condition, especially if made from hot-rolled bars. Cold-rolled bars are much stronger than the hot-rolled, but the best results from both types are obtained after heat-treatment. The heat-treatment generally given to this class of steel is to heat it to 1500 degrees and then quench, after which it is reheated to from 600 to 1300 degrees F. and permitted to cool down slowly.

The Heat-treatment of Steel Castings. — The Pennsylvania Railroad has for some time been experimenting on the effect of the special heat-treatment of steel castings, particularly bolsters and locomotive frames, and the results of these experiments were embodied in a paper presented at a meeting of the American Institute of Mining Engineers at New York on February 18, 1914. An abstract of this paper will be of material interest, as it indicates what may be done with steel castings when properly treated, thereby permitting in railway service greater strength of cast-steel parts without any increase in weight or space. The obscurity formerly surrounding the heat-treatment of steel has been for the most part removed by the development of our knowledge of the critical points of steel, pyrometers, furnace construction, and the testing of the finished product. The operations of the heat-treatment proper are taken up under the heads of 1. Heating for quenching; 2. Quenching; 3. Tempering.

Heating for Quenching. — Heating for quenching is best conducted slowly, especially in the case of castings of variable thickness. Cracks may occur either in heating or in cooling, due to different temperatures at different points of the casting. The castings should be thoroughly soaked at the maximum temperature (generally 1500 to 1600 degrees F.), one hour being sufficient for sections 12 inches in thickness. The minimum temperature which will produce the desired hardening effect will, in all cases, be found to be the most satisfactory, as the grain coarsens when the critical range is exceeded to too great an extent. All temperatures should be governed by a checked pyrometer with the hot junction to the heated object, and with several couples in a large furnace to insure a uniform temperature.

Quenching. — The casting should be transferred as quickly as possible from the furnace to the quenching bath, and in the case of large castings, such as locomotive frames, this is by no means a simple matter. The larger castings are best handled by means of cranes and rollers. The quenching agent employed is generally water or oil, preferably the former, because of its cheapness and cooling effect. With intricate castings it is generally best to use oil. With water it is possible to have a large tank and a large running stream, serving to maintain a uniform temperature. Castings should never be thrown in to rest on the bottom of the tank, but should be agitated to prevent the formation of a coating of vapor, retarding the quenching effect. It is also best, whenever possible, to quench the thicker portions first.

Tempering. — Whenever possible the tempering should be done in a bath of some kind, such as lead, barium chloride, a barium chloride-salt mixture, or oil. In the case of large castings this is manifestly impossible, and great care should be exercised in obtaining a uniform temperature in the tempering furnace. The use to which the casting is to be put determines the drawing temperature, railroad work, by reason of the shock and vibration of the road, requiring high ductility at the sacrifice of some strength; the temperature for this class of work, therefore, is about 900 degrees F.

Process for Hardening Cast Iron. — A patented process whereby cast iron in the rough or in the finished state may be hardened or tempered, the hardness extending completely through articles of comparatively large dimensions, is described in the following. One of the principal objects of this process is to provide a cheap and simple method of rendering iron castings so hard that they may be used for many purposes in the place of steel, thus reducing the manufacturing cost of a large number of articles. While subject of a patent, there is considerable doubt as to the practicability of the process.

The castings which are to be treated may be completely finished as regards machine work before they are hardened. The casting is first heated to a "cherry-red" heat; it is then dipped in a bath which consists of a practically anhydrous acid of high heat-conducting power, preferably sulphuric acid of a specific gravity of from 1.8 to 1.9, to which is added a suitable quantity of one or more of the heavy metals or their compounds — such, for example, as arsenic or the like. The preferable ingredients of the bath are sulphuric acid of a specific gravity of approximately 1.84 and red arsenic in the proportions of ¾ pound of red arsenic crystals to 1 gallon of sulphuric acid. The castings may be either suddenly dipped in the afore-mentioned mixture and then taken out and cooled in water, or they may be left in the bath until cool. We find, however, that dipping the castings in the bath and holding them there for a time which varies according to the 'size of the castings, and then completely cooHng in water, is quite as satisfactory and produces a material which is just as hard as if the castings were allowed to remain in the bath until cool, and the former method is preferable if a large number of castings are to be hardened, as the bath is thus prevented from becoming overheated. In preparing the bath when sulphuric acid and red arsenic are used, we find that better results are obtained when the crystals are added to the sulphuric acid and the bath is allowed to stand for about a week before using, the reason probably being that the bath becomes more saturated with arsenic compound when the dissolved red arsenic has been long in the sulphuric acid.

It is not necessary that the casting be machined completely before hardening, as rough-finished castings may be hardened equally well. The change which takes place in the metal is in the nature of a molecular re-arrangement or re-crystallization coincident with an increase in the combined carbon at .the expense of graphitic carbon.

It is found that the more rapid the cooling of the metal, the harder it will become. For this reason the bath must be of high heat-conducting power, and this requirement is obtained by the use of the ingredients referred to. Furthermore, the bath must be practically free from water, as it is found that when the acid contains water in any considerable quantity, a steam cushion is formed between the acid and the metal which prevents their coming in contact, with the result that the cooling is less rapid, and, consequently, the iron is not so hard.

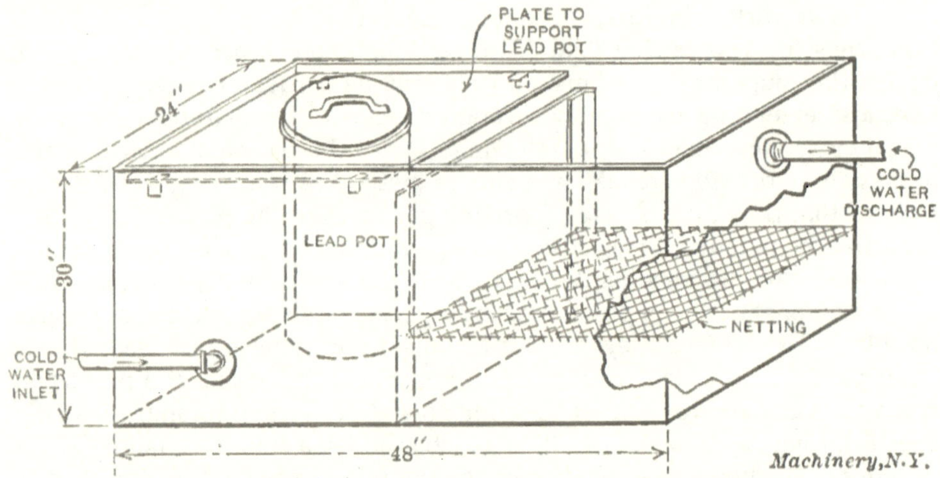

Fig. 9. Cooling Tank and Hardening Bath for Cast Iron

A cylindrical jar made of lead should contain the bath, the size varying according to the work to be done. A jar about 10 inches in diameter by 18 inches deep will be about fight for ordinary small work. This lead jar should be enclosed in an outer vessel through which water is caused to continuously circulate in order to keep it cool. It might be further pointed out that if it is desired to harden one portion of a casting and leave the remaining portion soft, this may be accomplished very readily by immersing only the part to be hardened.

Such a hardening equipment has proved very satisfactory in locomotive work for hardening bushings, etc., and many other uses could no doubt be found for it. The whole equipment can be homemade, consisting as it does simply of a steel tank divided into two compartments, and the lead jar for the bath, as shown in Fig. 9. The water circulates in the first chamber around the lead vessel, keeping it cool, and then passes to the second chamber, into which the castings are dropped, when taken from the bath, to cool off. The water then passes into the sewer or other suitable containing device. There is a screen placed in the second compartment to keep the castings from falling to the bottom.

Local Hardening. — One method of hardening locally is to cover the part that is to remain soft with a thin metal shield, so that it prevents the surface from being suddenly cooled by the direct action of the cooling medium. The steam or vapor which forms beneath the cover prevents the cooling medium from entering until the work has cooled sufficiently to prevent hardening; hence, a rather loose-fitting shield is desirable. The shield should be made of sheet iron or steel of about No. 29 gage (0.014 inch), for ordinary work. It is composed of one or more pieces, depending upon the shape of the part, and, when several pieces are required, they can be bound together with wires or rivets. Of course, the surfaces to be hardened are left exposed. The heating should be done in a furnace or open-forge fire. A lead bath should not be used, because the hot lead beneath the shield will cause an explosion when the part is cooled. The quenching bath can be the same as when the shield is not used.

Local hardening is also effected by the application of a compound called "Enamelite" to the parts which are to remain soft. This compound, for tool steel, is in the form of a powder which is mixed with hot water to form a paste. It has the property of clinging to the steel and liberating hydrogen (the greatest known non-conductor) when the heated steel is plunged into water. This causes the steel to retain its heat long enough to escape the chill, so that it remains soft where the enamelite has been applied.

Chapter Ten - Casehardening

General Principles of Casehardening. — Casehardening is the process of increasing the carbon content of the surface of steel comparatively low in carbon, so that it can be hardened by the usual method of being heated to the hardening temperature and quenched in a cooling medium. The term casehardening, by itself, implies the hardening of the surface or skin of an article, and in order to fully understand the process and its object, it is necessary to briefly consider the facts and laws upon which it is founded. Carbon has a very great affinity for iron and combines with it at all temperatures above a faint red heat. Advantage was taken of this fact in the production of steel by cementation, an old process which consisted of rolling wrought iron into thin strips and then placing these in boxes with some material containing a fair proportion of carbon, These boxes were then heated to a very high temperature and the carbon was gradually absorbed by the iron.

The process of casehardening is, in fact, only an improvement on this old cementation process used in times past for making steel from wrought iron. The steel is heated in packing boxes in the presence of a carbonaceous material and when the surface of the steel has absorbed enough carbon so that it will harden the same as high-carbon steel, it can be quenched in oil or water, according to the requirements. For many purposes, in machine work, articles are required which must have a perfectly hard surface and yet be of such internal structure that there is no chance of breaking them when in use. In many instances, this result can be obtained better by using casehardened mild steel than by using high-class crucible steel. For example, in making axles, cups, cones, and many similar parts for bicycles, it is extremely difficult to obtain perfect hardness combined with great resistance to torsional, shearing or bending stresses. For such purposes, nothing meets these requirements so well as do articles which have been casehardened.

A great improvement has been made in casehardening processes during the last few years. The advance was begun with the development of the bicycle industry; and the necessity for casehardened parts of the highest quality in automobile manufacture has caused a still further improvement in this field.

As an example of what has been accomplished by proper casehardening method's, consider the transmission gearing of an automobile. Who would think of throwing in the back-gears of a lathe or any other machine tool without first stopping the machine? In an automobile, however, this very thing is actually done dozens of times a day, by a person who gives little thought to what he is really doing. Yet the gears stand up under this treatment because of being manufactured of special steels developed during recent years and because of being heat-treated and casehardened by improved methods.

There are a number of different questions that must be considered in order to obtain good results in casehardening. In the first place, the proper kind of steel to be used for various purposes must be carefully selected. Another most essential thing is that the casehardening furnace must give a uniform heat. As oil and gas have to a great extent superseded coal as fuels for casehardening furnaces, the changes in furnace construction have, of late, been considerable. Another item which must be given careful consideration is the box in which the material is packed, as well as the carbonaceous material itself used in packing the parts to be casehardened. Still another question to be dealt with is the method used for hardening the parts after they have been carbonized.

Steel to be used for Casehardened Parts. — As the casehardening process consists in adding carbon to the steel, a material must be used which will absorb carbon without necessitating overheating or burning. The effect of carbon on steel is, in general, it may be said, to make it dense, and the denser the steel the higher the heat necessary to open the pores through which it must absorb the carbon. A low-carbon steel containing, say, from 0.15 to 0.20 per cent of carbon is, therefore, most suitable for casehardening. It should also be borne in mind when selecting the material that the casehardening process does not eliminate any of the impurities ordinarily found in iron, such as sulphur, phosphorus, etc., and, hence, a material as free as possible from these impurities should be selected; besides, the material should be perfectly sound and free from mechanical faults or weaknesses caused by overheating or improper working during the manufacturing processes.

Both iron and mild steel have been employed as materials for casehardening in the past; but this is the steel age, and iron has long passed its day. The steel employed should be prepared, selected and controlled from the beginning with the object of making it suitable for the final requirements. There are many points with relation to the selection of the proper steel, its composition and treatment, which can only be gained by long experience and a study of the requirements, but, as a general rule, the low-carbon steel specified in the preceding paragraph will be found suitable for most purposes.

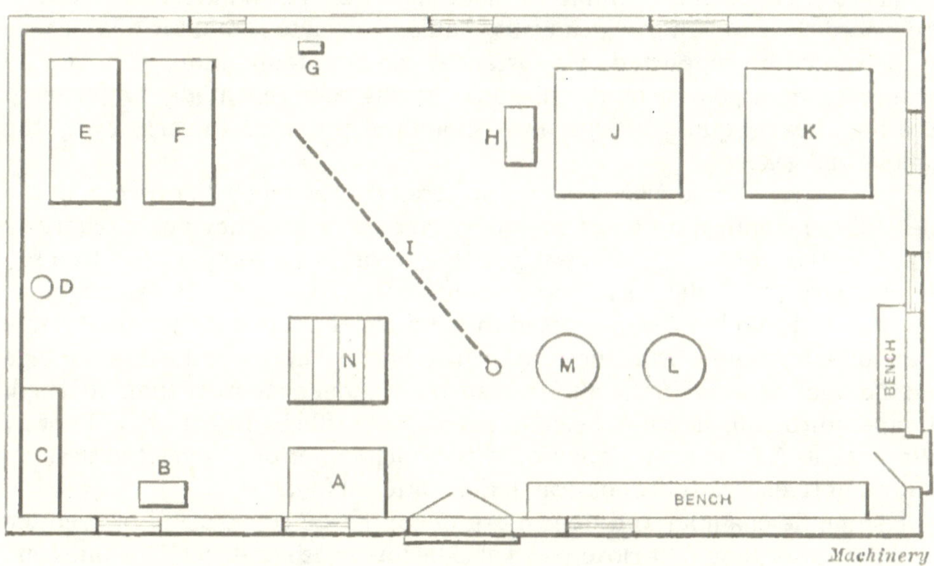

Fig. 1. Plan of a Well-equipped Hardening Room

Hardening Room Equipment. — In Fig. 1 is shown a plan of a hardening room especially well equipped for its purpose, and containing the requisite furnaces and appliances for casehardening. This illustration shows the arrangement of the heat-treatment equipment of the Boston Gear Works, which is located in a separate fire-proof building 26 by 45 feet, consisting of a steel framework covered with asbestos-protected metal. The equipment comprises two carbonizing furnaces *J* and *K* using coal as fuel and having heating chambers 24 by 60 inches, and also the following oil-burning furnaces: one Westmacott furnace, 10 by 12 inches, located at *B*; one Rockwell furnace, 24 by 36 inches, located at *F*; one American Gas Company's furnace, 20 by 60 inches, located at *E*; one Stewart crucible furnace, located at *D*. At *A* is a lead reheating furnace. Besides the furnaces the equipment is as follows: *H* indicates a forge; *I*, a jib crane; *L*, a water-quenching tank; *M* and *N*, oil-quenching tanks; and *C*, a bin for carbonizing materials. The temperatures are indicated by a Bristol thermo-electric pyrometer, located at *G*. The steel jib crane allows heavy work to be easily placed in or removed from the furnaces and cooling tanks.

The Casehardening Furnaces. — In building or constructing a furnace for casehardening, the size of the work to be casehardened should be the first consideration. It is far better to use a small furnace with a small box, whenever possible, than to make the furnace larger than is absolutely required. If the work varies in size, a number of furnaces of different sizes may be used. Small furnaces require less fuel, and small work should be placed in small boxes, as otherwise the pieces packed near the sides will be overheated before those in the center will reach the required temperature. When several furnaces are used, these should be made right- and left-hand so that two can be placed close together. Thick walls should be used to retain the heat. These walls should be supported by substantial concrete foundations so that they will retain their position and shape after having been repeatedly subjected to the high heat required. Large flues should be provided to carry away the smoke and gases.

The furnace should also be so constructed that as much as possible of the heat of the combustion gases may be extracted before they are discharged. The construction should make it possible to raise the temperature to a full orange heat (1830 degrees F.) and maintain it at this heat with fair regularity. It should also be so constructed that neither the fuel nor the direct flame can come in contact with the charge. The flames should uniformly impinge on the sides and roof of the muffle in such a manner as to raise them to a high temperature, thus heating the contents of the muffle by radiation and not by direct heat. A furnace designed on this principle not only gives the best results, but is also most economical in the matter of fuel.

The muffle chamber and flues must, of course, be constructed of firebrick and the doors should fit closely and also be lined with firebrick. The flues and all parts of the furnace should be easily accessible. The door should be the full width of the oven so that the tiles can be taken out and the flues cleaned. It is important that there should be a small peep-hole in the door with a cover plate. A hole 1½ inch in diameter is quite large enough. This latter is a most important detail as it provides against the need of opening the door in order to judge the heat and furnishes the most accurate means of estimating the temperature by the eye. The furnace must also be fitted with a reliable damper plate or other effectual means of controlling the draft.

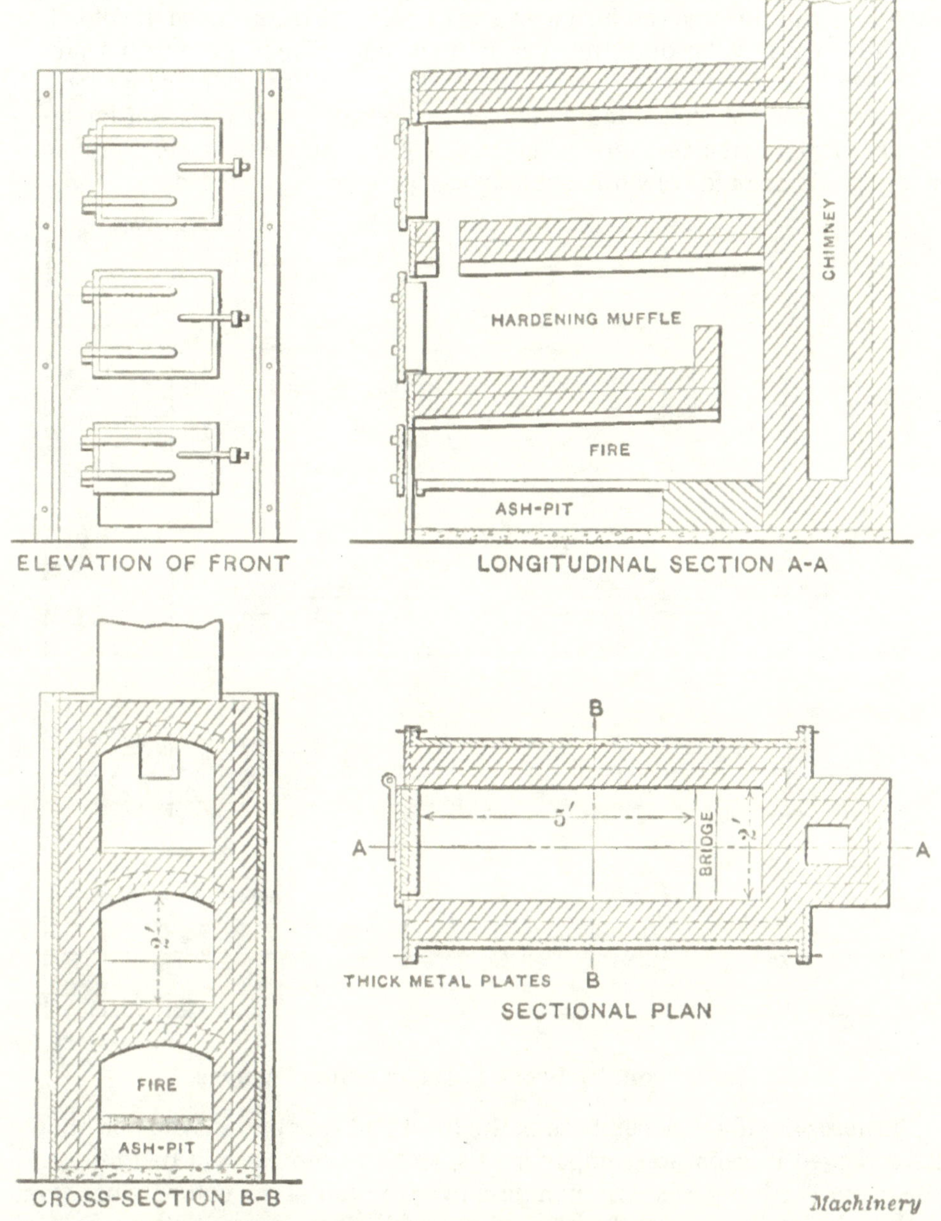

Fig. 2. Plan and Elevation of Casehardening Furnace

Oil-burning Furnaces. — In an oil-heated furnace, a light oil having a high heating value and comparatively free from carbon deposits should be used. All piping should preferably be placed above the furnace so as to be easily accessible. If, however, the piping is placed below the ground, it should be arranged in compartments which can be easily reached, if repairs are required. A pressure blower should be used and the pipes from the blower should be run through the furnace so as to pre-heat the air used. If cold air is used directly, it will reduce the heat in the furnace. The furnace fronts should be made in several parts to prevent cracking, with the door properly balanced and lined. A shelf should be provided projecting at the front for holding the boxes when they are taken out or put into the furnace. The smoke stack should be made of sufficient height to produce a good draft.

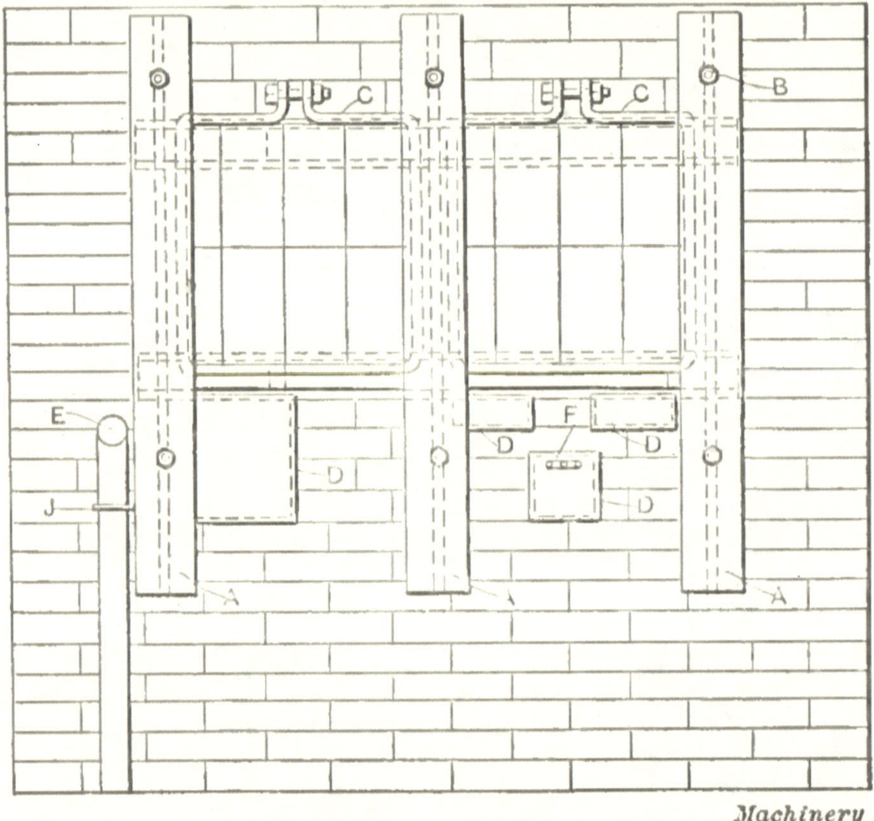

Fig. 3. Front View of a Casehardening Furnace

Burners should be placed both at the front and rear of the oven and should be arranged in separate compartments, so that the heat will be uniform in the oven. The hot gases will then pass over the top of the compartment wall and strike the boxes on the top, after which they pass out through small openings in the corner of the furnace. They then take a zigzag course under the tiles and pass from there through a flue to the rear of the furnace. A large conduit should be provided just below the ground which will catch all the soot. This conduit should be provided with iron covers which can easily be taken off to remove the accumulation of soot.

The furnace should not be heated too quickly, as this is apt to crack the brickwork. The cooling should also be done gradually. After the work has been taken out and the heat shut off for the day, all the dampers should be closed to hold the heat. In this way the furnace will cool slowly and cracking or bulging out of shape will be prevented. In addition, it will be easier to heat the furnace the next morning, as it will have retained some of the heat.

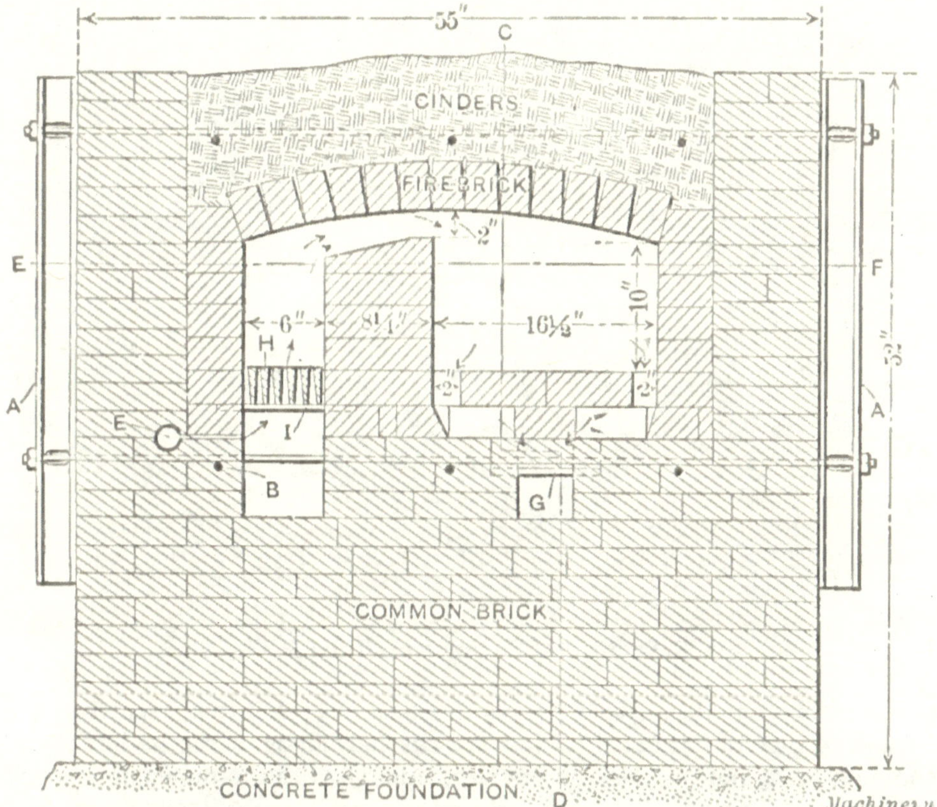

Fig. 4. Section of Casehardening Furnace on line *XY*, Fig. 5

Types of Coal-burning Casehardening Furnaces. — In Fig. 2 is shown a diagrammatical view of a furnace which may be found useful as a guide for the erection of furnaces using solid fuel. The upper chamber in this furnace is not necessary for casehardening, but it may be found useful to have such a chamber and to employ it for annealing small articles while casehardening is being done. This will add only slightly to the amount of fuel used. In furnaces not having this upper chamber, work to be annealed may be placed in the furnace after the work to be casehardened has been removed and then the furnace may be brought to the proper heat. The material to be annealed can then remain in the furnace until the next morning with the drafts closed up and the fire banked.

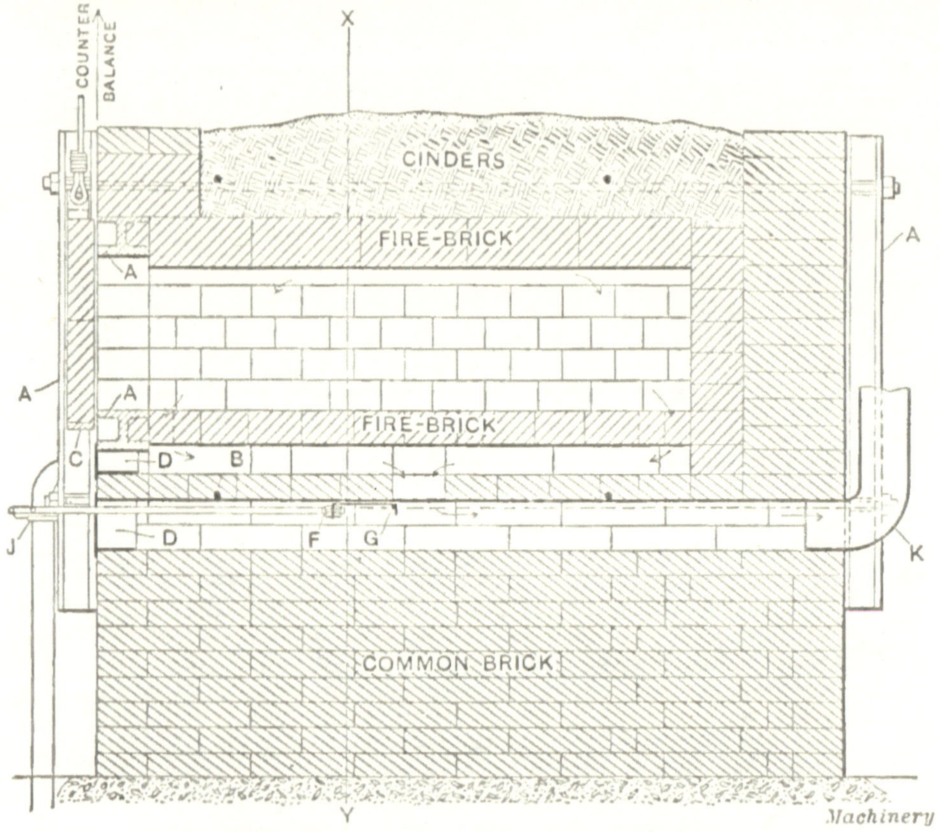

Fig. 5. Section of Furnace on line *CD*, Fig. 4

The casehardening furnace, shown in detail in Figs. 3 to 6, inclusive, is a very good type of hard coal furnace for casehardening. It can be built from common brick and firebrick and is large enough with the dimensions given for ordinary shop work.

If a larger size of furnace is required, it will be necessary to use large tile in place of the firebrick for the bottom of the oven. A blast is used in connection with this furnace when starting the fire, but very little artificial draft will be required after the box containing the work to be hardened is red hot; this, of course, depends to some extent upon the draft in the chimney. A damper is supplied in a pipe behind the furnace to regulate the heat. The following are the principal parts of the furnace: *A*, cast-iron stays; *B*, 5/8-inch staybolts; *C*, door frame, ¾ by 1¾-inch iron; *D*, sheet-iron caps for flues; *E*, blast pipe, 1½ -inch gas pipe; *F*, damper; *G*, damper support; *H*, cast-iron grate; *I*, grate support; *J*, blast shut-off; and *K*, smokestack connection.

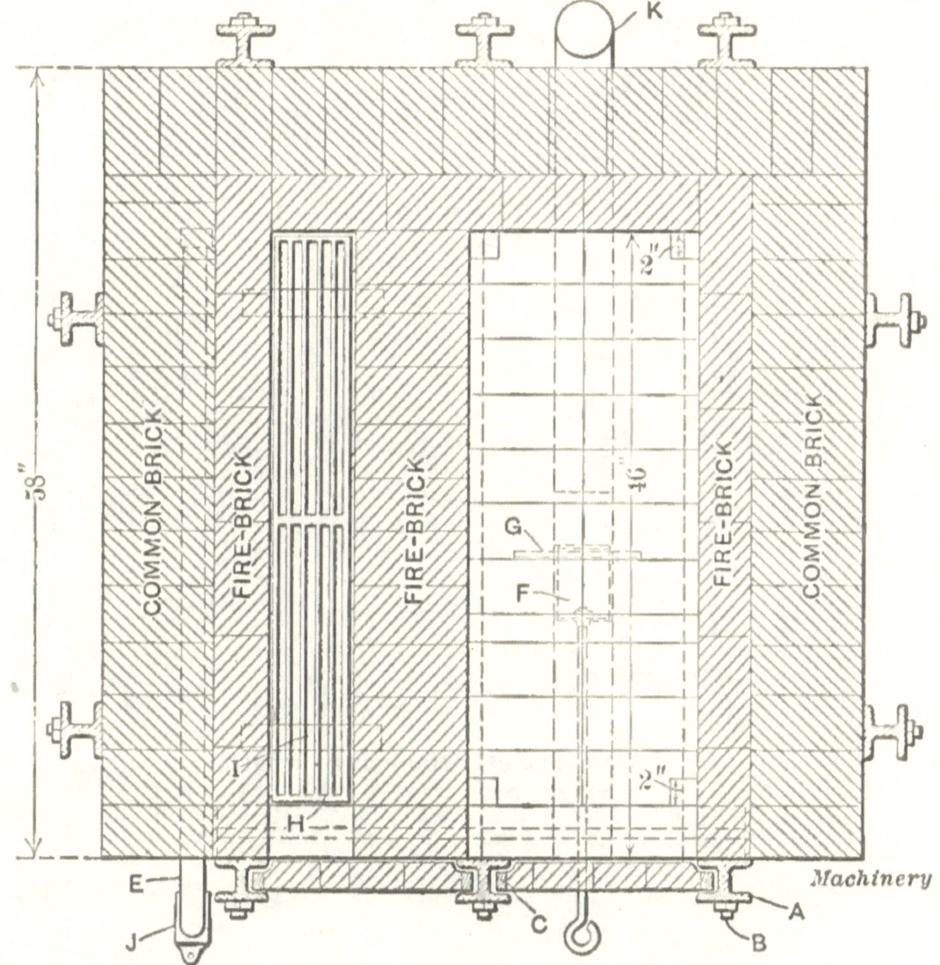

Fig. 6. Section of Furnace on line *EF*, Fig. 4

Figs. 7 and 8 show the type of casehardening furnace used by the Boston Gear Works. This furnace is built of brick and is of the same general type as the furnace just described. It is a purely reverberatory furnace, the heat being deflected from the firebox *A* over a bridge *B* down upon the work, the firebox *A* and furnace chamber *C* being arched over. The gases from the furnace chamber C descend through four flues *D*. These flues unite in two ducts *E* which lead to the chimney flue *F*.

Boxes for Casehardening. — Boxes for casehardening should not be made larger than is necessary' for the class of work being handled. They are made both from cast and wrought iron, the former being cheaper in first cost, but the latter withstanding reheating so many times that they are cheaper in the end. Cast iron is also objectionable because it is porous and seems to absorb carbon from the carbonizing material. Very good boxes can be made

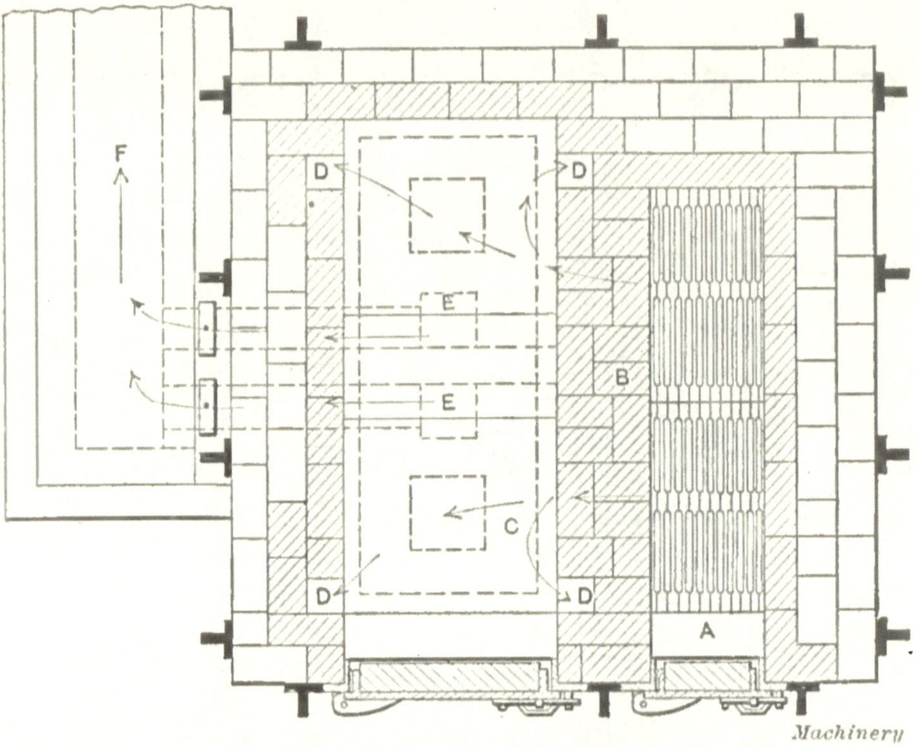

Fig. 7. Plan View of Carbonizing Furnace

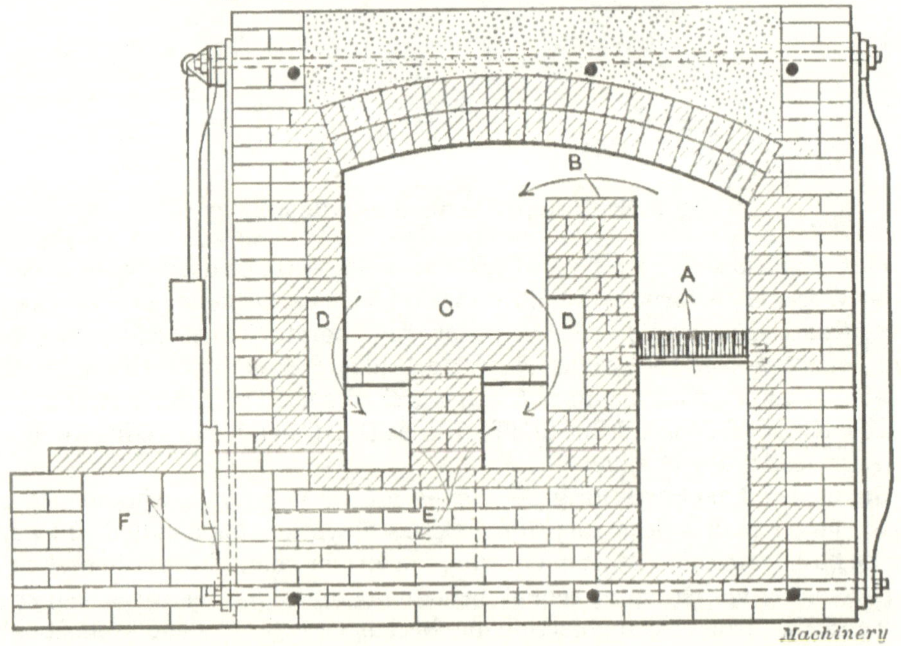

Fig. 8. Sectional Elevation of Furnace shown in Plan View in Fig. 7

from malleable iron. The precaution of not making them of too large dimensions is quite important, as otherwise there is a great risk of the articles in the middle of the charge not being carbonized to a sufficient depth on account of the heat being low. For such parts as bicycle axles, pedal pins, and the like, the box should not be larger than 18 by 12 by 11 inches, while for small articles like cups, cones, etc., for bicycles, 12 by 10 by 8 inches is large enough. The box should have a cover fitting closely on the inside of the box.

The boxes should be provided with feet, as shown in Fig. 9, so that the heat can circulate all around them. The covers should be provided with ribs on the top to prevent excessive warping, and the sides of the box should be ribbed so that a fork or grapple iron, such as shown in the upper part of Fig. 10, can be used for handling the box. The sides of the boxes should taper slightly towards the bottom so that the contents can easily be dumped out of them. It is also easier to cast the boxes when made in this way. A simple truck for handling heavy boxes is shown in the lower part of Fig. 10.

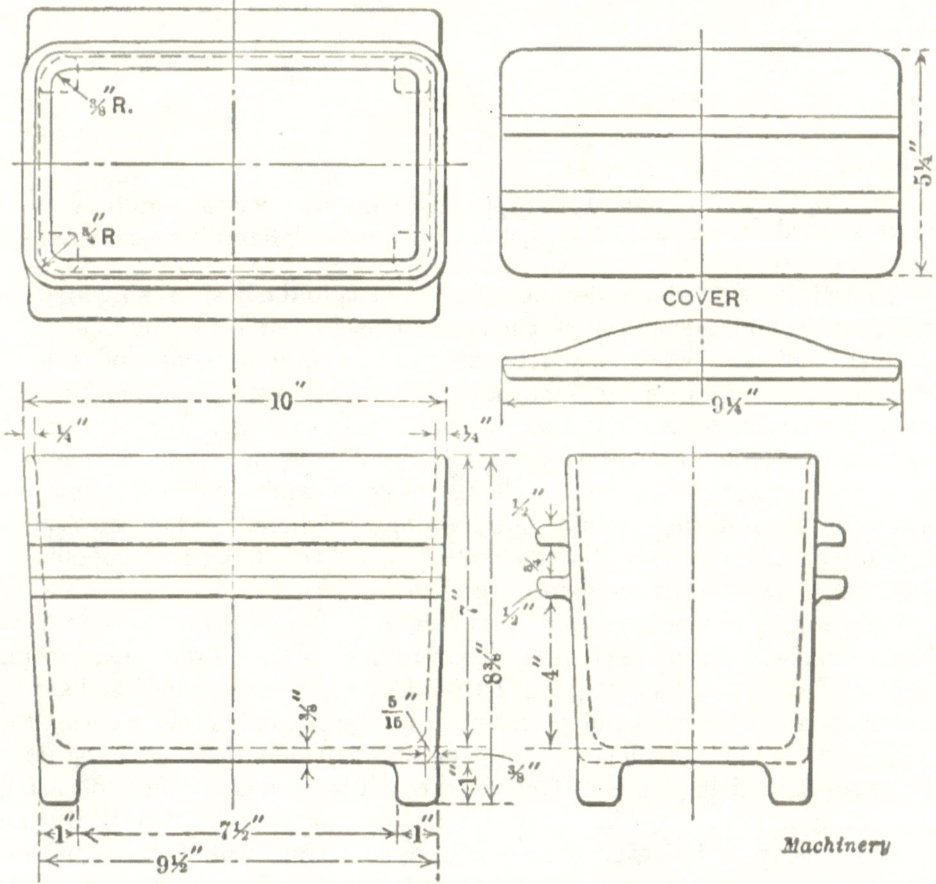

Fig. 9. Box of Approved Design used for Casehardening

When very large boxes are required, they should, if possible, be provided with a hole through the center, so that the heat can reach the contents from the inside as well as from the outside. A box of this kind is shown in Fig. 11. For long work, such as shafts, tubing, etc., a wrought-iron pipe with a cap on each end forms an ideal casehardening box. As a general rule, the boxes should not be made too deep in proportion to their other dimensions, as it makes it more difficult to pack the parts into them if made in this way.

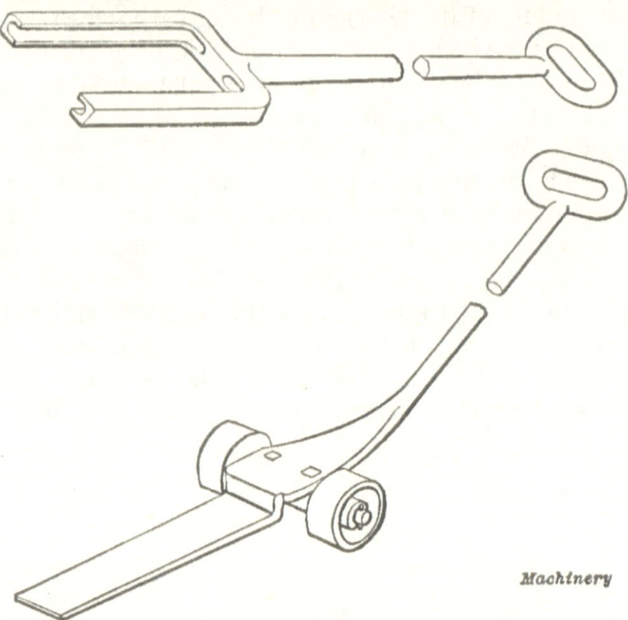

Fig. 10. Grapple Iron or Fork used for Handling Casehardening Box, and Truck for Handling Heavy Boxes

Packing Materials used in Casehardening. — There is considerable difference of opinion as to the best packing materials to be used in casehardening. The carbonizing materials in general use are charred leather, powdered bone, cyanide of potassium, wood and animal charcoal, prussiate of potash and other materials consisting of mixtures of carbonaceous materials and certain cyanides and nitrates. For very slight hardening, cyanides are often used, but no great depth of case should ever be attempted with these. Charred leather gives good results, but poorly charred leather and that made from old boots, belting, etc., should not be used. A mixture preferred by some to charred leather consists of 60 parts of wood charcoal and 40 parts of barium carbonate.

Theoretically, the perfect carbonizer should be a pure form of carbon. This being impossible in practice, some claim to have obtained the most certain and satisfactory results with good charred leather, whereas others advise the use of good fine charcoal, pulverized to about the size of kernels of corn. Others, again, claim that while there are a great many different kinds of hardening materials that give satisfaction, the old-fashioned method of using ground bone can always be relied upon to give satisfactory results. During the last few years, however, the use of bone in various manufactures has increased so that the price of ground bone for casehardening purposes is almost prohibitive. This has caused leather to become more and more extensively used for this purpose, the leather being first burned, and then ground and graded. The carbonizer should always be perfectly dry and in a granulated or powdered form.

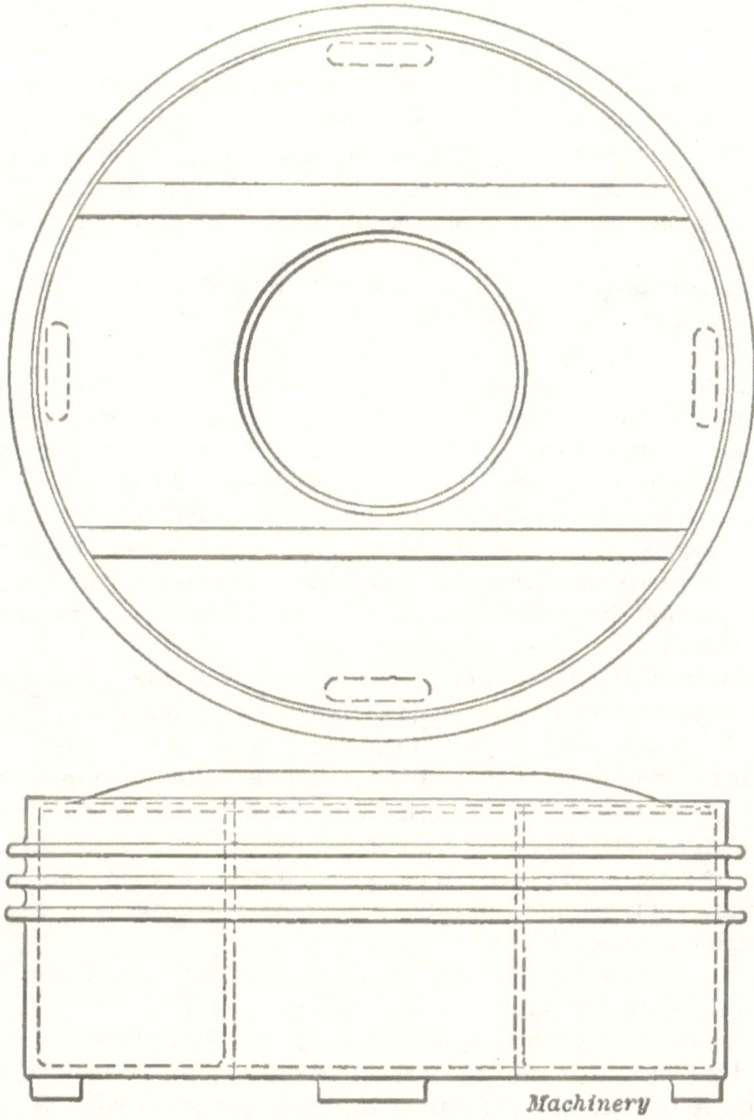

Fig. 11. Large Circular Box with Hole in Center for the Circulation of the Gases of Combustion

A mixing bin is of great advantage in connection with the handling of casehardening material. Some partly used bone and some new is then used to make a mixture suitable for the size of the pieces to be hardened. Large pieces require a richer material than smaller ones, because during the higher heat required for the larger pieces and the longer application of the heat, more carbonizing material will burn away.

The packing room should, if possible, be separate from the room containing the furnaces, so that the packing can be done without the discomfort of the heat and dust. Tables on wheels, or trucks, provided with shelves of the same height as the shelf in front of the furnace and large enough to hold the required number of boxes for one furnace, should be provided, so that the packed boxes can be easily moved to the furnace and quickly placed in it. The work to be hardened should be classified according to its size and the percentage of carbon required, as it will take a higher heat for larger work, as well as for pieces which are required to absorb a higher percentage of carbon.

Packing the Boxes. — When packing a box, first put a layer of the casehardening material on the bottom, the thickness of this layer depending on the size of the pieces to be hardened. About 1½ inch is the minimum depth that should be used. This layer should be well pressed down and upon this is placed the articles to be hardened. Care must be taken to leave sufficient space all around each piece to prevent the parts from touching each other or the walls of the box. A space of 1½ inch should be sufficient. If the articles are heavy they do not require such great care in packing, but if they are thin or long, or have a peculiar shape, greater care is required so that the pieces are properly supported and cannot sag out of shape. Thin, long pieces should, if possible, be placed in an upright position in order to prevent trouble from this cause.

It has been claimed that the precaution of preventing one piece from touching another is unnecessary, and that no harm is done if parts should be in contact with one another. This, however, does not seem a safe method to pursue, because in order that the proper case of high-carbon metal may be provided, all the surfaces of the parts should be in contact with the carbonaceous material, and if the parts touch one another, it is likely that they will be softer at the spots where they are in contact. It has also been found from experience that if there is not enough of the carbonizing material in the box, the work is liable to have soft spots.

When the first layer of work has been placed in the box, another layer of carbonizing material is put over it and well pressed down, care being taken not to displace any of the articles already packed. Another layer of parts to be hardened is then laid on the carbonizer, and this is continued until the box is nearly full. About two inches from the top of the box sheet-steel strips about 1/16 inch thick should be laid on the last layer of carbonizing material and these, in turn, should be covered with a layer of about 1 inch or more of powdered charcoal. Then the cover is placed on the box and the edges are sealed with fireclay or asbestos cement. The clay used for luting around the cover, and which is also used for "stopping off" portions to be left soft, must be of good quality and free from grease. Clay contaminated with grease in any way will cause irregularity in the product obtained. The contents in the box should be packed as compactly as possible, because the more solidly the box is packed the more complete is the exclusion of air.

Carbonizing in the Furnace. — The heat required for carbonizing is a great deal higher than that required for ordinary hardening. If, for example, the material to be casehardened were heated only to 1400 degrees F., which would be sufficient for the hardening of ordinary tool steel, the result would be very unsatisfactory; in fact, there would be no result at all. Small parts must be heated to at least 1575 degrees F., in which case sufficient depth of carbonized surface will be obtained in from six to eight hours. The time given as correct for casehardening should be taken from the time the boxes are heated clear through. Ordinarily, however, the proper heat for casehardening is about 1800 degrees F. for a full orange heat, and this should be maintained with great regularity throughout the operation.

The length of time occupied in carbonizing is regulated by the depth of casing required and also by the dimensions of the article. At the close of the carbonizing period, the box is withdrawn from the furnace and placed in a dry place where it is allowed to become quite cold. It is then opened, the articles taken out and brushed over to remove all adhering matter. If the box has been properly packed and luted, the articles should be quite white, or at least have only a slight film of deep blue color. The denser and more inclined to redness the surface is, the more imperfect has been the packing and sealing of the box.

If there is any doubt about the length of time required for heating the pieces to obtain a certain depth of case, wire a couple of pieces together, allowing the wire to project out of the box. These pieces can then be taken out quickly and hardened, and in this way it can be ascertained whether the parts have been sufficiently carbonized. In casehardening very small work, it is advisable to wire the pieces together so that they can be taken out of the box at once; otherwise, they would have to be picked out with small tongs, as it is impracticable to sift very small work in a sieve because the mesh would have to be so fine that it would take a long time to do the sifting.

The boxes should never be put into the furnace under a high heat, but should be placed in it when its temperature is from 800 to 900 degrees F. Then the heat should be slowly brought up to from 1500 to 1800 degrees F. In placing the boxes in the furnace, great care should be taken that the hot gases have an opportunity to circulate all around them. A pyrometer should be put in some convenient place and properly wired so that the heat in the furnace can be readily ascertained at any time. If there is a great deal of night work to be done, a recording pyrometer should be used as it gives the man in charge a record of the heats during the night.

By the aid of the pyrometer it has been found that when an oil or gas furnace is used it is necessary to have an expansion tank in order to get a constant air pressure, otherwise the pulsation from the blower will affect the heat in the furnace. This expansion tank should be situated so that the blower is connected directly with one end, while the discharge pipe is connected at the opposite end. This will then act as a reservoir, producing a constant pressure. When oil is used for the heating, it is preferable to pump it from the storage tank in the ground to a stand pipe, which will insure a constant flow of the oil. The intermittent action of the pump, should the oil be used directly as it comes from it, is objectionable. There is also another advantage, in case the pump should have to be shut down on account of break-down. In that case, the furnaces could still continue to operate, as the stand pipe should hold a supply of oil sufficient for several hours. At night and on holidays the oil should be drained back into the storage tank in order to minimize the danger incident to its use.

The supply pipe for the air should come from the outside and should be so arranged that the air passes through a tine wire netting, so as to prevent foreign substances from entering the blower.

On the outside of each furnace a card should be placed telling the kind of work that is in the furnace, when the work was put in, the heat required for it, and when it is to be removed. These cards can be kept as a record which will be of value when comparison is made with the depth of case obtained under any specific conditions.

Experiments to Ascertain the Proper Carbonizing Temperature. — Although the proper temperature for casehardening is about 1830 degrees F., this temperature may be modified to suit the purpose in view. The absorption of the carbon commences when the steel reaches a low cherry-red heat (1300 degrees F.); it begins, of course, at the outer surface and gradually spreads until the whole of the steel is carbonized. The length of time this requires depends upon the thickness of the metal being treated. The percentage of carbon absorbed is governed by the temperature, and although the increase of carbon is not in uniform proportion to the rising temperature throughout, it is perhaps sufficient for our present purpose to note that at 1300 degrees F., iron, if completely saturated, can contain no more than about 0.50 per cent carbon; at 1650 degrees F., about 1.5 per cent carbon; and at 2000 degrees F., about 2.5 per cent. These results, however, are only obtainable when the whole section of the iron has received all the carbon it is capable of absorbing at the given temperature, and is therefore in a state of equilibrium. From this it will be seen that if the process is stopped before the action is complete, the central parts of the iron must contain less carbon than the outside, and upon this fact the process of casehardening is founded.

If we take two pieces of 5/8 inch diameter round mild steel, and heat one of them with a carbonizer at a cherry-red heat, and the other at a bright orange heat, for six hours, the first will be cased to a depth of about 1/32 inch, and the other to a depth of nearly 1/16 inch, while the amount of carbon taken up will be about 0.50 and 0.80 per cent, respectively; so that, so far as regards the hardness of the skin, the piece carbonized at the higher temperature gives the best result. From this we learn that a temperature of 1830 degrees F. will give us sufficient hardness of case.

We have next to find which temperature has the least harmful effect on the mild steel core, and this can best be found by heating pieces of the mild steel at varying temperatures at and above the selected one for the same length of time, using lime or other inert substance in the pot instead of a carbonizing material, and afterward reheating and quenching in water. Suppose, for example, we take three pieces, heating at 1830, 2370 and 2730 degrees F., or full orange, white and bright white respectively. We shall find that those at 2370 and 2730 degrees break very short and have lost nearly all their original tenacity, while that at 1830 degrees appears tougher and altogether stronger than before.

Having arrived at a knowledge of the right temperature, it remains now to inquire as to the length of time requisite to yield a sufficient depth of case. At a full orange heat a bracket cup of ordinary dimensions should in two hours be hardened 1/32 inch deep, and a bracket axle of 11/16 inch diameter in 6 hours would have a case 1/16 inch deep. From this it will be seen that the speed of penetration is not in exact proportion to the time of heating.

Variations in Casehardening Methods. — The correct way in which to caseharden is first to carbonize the material and then to allow the boxes to cool down with the work in them, after which they are reheated and hardened in water. The reheating refines the grain of the steel and prevents the formation of a distinct line between the outer hardened case and the soft core. If there is a distinct line between these two sections, the case is liable to flake off when the hardened part is subjected to severe stresses.

A still more refined method of casehardening is to repack the work, after it has been carbonized, in old bone, and after heating for two or three hours take it out and dip the pieces in the hardening tank directly as they come from the boxes. This will produce a very fine grain and in many cases prevent warping. If the work is large and it is required to toughen the inner core, it should be reheated to a higher heat than otherwise; then, after dipping, reheat again to 1500 or 1600 degrees F. according to the size of the work, and redip.

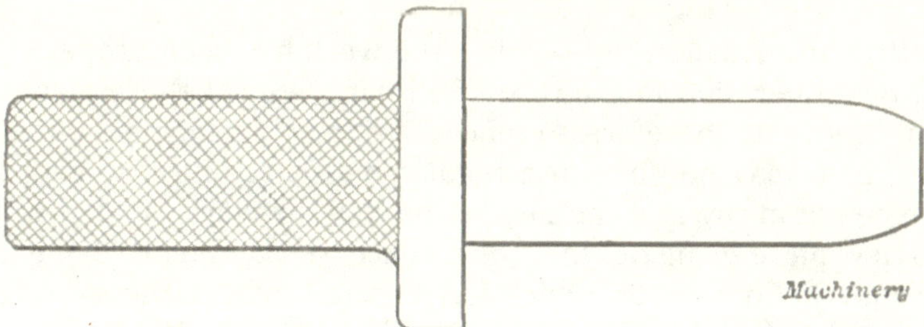

Fig. 12. Mandrel used when Hardening Collars and Similar Parts on the Outside only

However, if the work to be hardened consists of bolts, nuts, screws, etc., it is satisfactory to dump them into water directly from the furnaces, without any reheating. A regular iron wheelbarrow with two pieces of flat iron placed across it lengthwise should be provided. On top of these bars is placed a sieve made from 1/8-inch wire with ¼-inch mesh, about 18 inches square by 6 inches deep. This sieve should have a handle 6 feet long and 5/8 inch in diameter. The boxes are emptied into this sieve, and after sifting, the heated material is dumped into a tank of cold water which should be of sufficient size to prevent the water from heating too quickly. Care should also be taken in emptying the contents of the boxes into the water that they are not all dumped in one place, but scattered about in the tank. A constant flow of water should be available while the work is being hardened. The work should under no circumstances be removed from the furnace until the heat has been lowered, as the steel should be treated as tool steel after it is carbonized, and it would be injurious to the steel to harden it at the high carbonizing heat.

Gears and other parts which should be tough, but not glass hard, should preferably be hardened in an oil bath. There is then less liability of warping the work, and the hardened product will stand shocks and severe stresses without breakage. Cotton-seed oil is the best hardening medium to be used in this case.

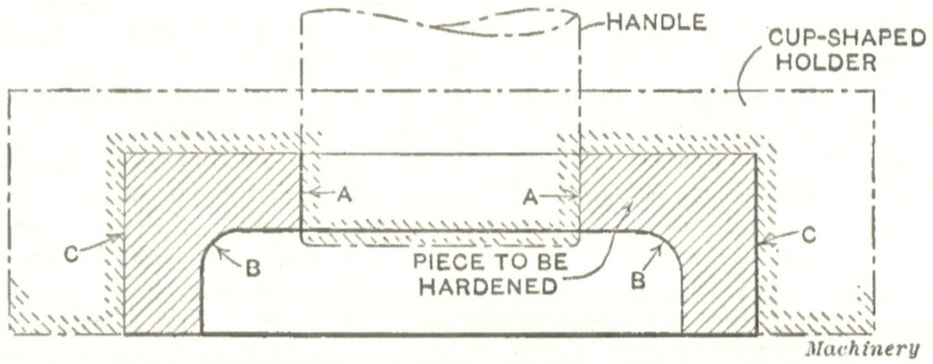

Fig. 13. Holder for Parts which are to be Hardened Locally

Reheating and Hardening. — After the work has been properly carbonized, the next operation in the case of all parts, except those mentioned as exceptions above, is, therefore, to reheat. This may be done either as already explained, or it may be done in a regular muffle gas furnace in which the work can be put in rows on the tile. In this way the work can be heated very slowly, a new piece being put into the furnace to take the place of each piece as it is removed. Collars, etc., which are required to be hardened on the outside, but ought to be left soft on the inside, should be hardened on a mandrel, such as shown in Fig. 12, the diameter of the mandrel being from 0.001 to 0.003 inch smaller than the hole in the piece to be hardened. If the inside of the piece only needs to be hardened and the outside should be left soft, a cup-shaped holder, such as shown by the dash-dotted lines in Fig. 13, may be used. In this case the work will harden at *B* while it is left soft at *A* and *C*.

Reheating Furnaces. — It is important that a suitable furnace should be employed for the purpose of reheating casehardened articles before quenching. It is not advisable to reheat in the same furnace as is used for carbonizing unless the furnace is run specially for this purpose and at a lower heat. Ordinarily, a small gas or oil-heated furnace can be used to advantage for reheating. A properly constructed gas or oil-heated furnace can be regulated with great exactness and this is very important in all hardening.

Quenching Baths. — It is generally advisable when quenching casehardened parts to use brine or salt water rather than pure water, because the latter does not chill the parts quickly enough. The hardening tank should be about 30 inches in diameter and 36 inches deep and have a constant flow of water from a pipe in the center about 6 inches below the surface. When extreme hardness of case is not required, pure water will, of course, give satisfactory results, and where the hardness of the surface can be still further sacrificed for a tenacious structure in the material, oil quenching baths are frequently used.

American Society for Testing Materials Methods of Casehardening. — A committee of the American Society for Testing Materials recommended the following practice for casehardening carbon-steel parts. Four different conditions were considered, varying from the heat-treatment that would give the hardest surface and the least strength, to that which would give the greatest strength with the least hardness of surface. When a hard case is the only requirement and lack of toughness or even brittleness is unimportant, the articles may be quenched by emptying the contents of the casehardening boxes directly into cold water or oil. In this way both the core and the case are coarsely crystallized and the strength is reduced. If the articles are allowed to cool to a temperature slightly exceeding the critical range of the casehardening, usually from 800 to 825 degrees C. (1472 to 1517 degrees F.), and then quenched, the core and case still remain crystalline, but the danger of distortion or cracking in the quenching bath is reduced and the strength is somewhat increased. The next recommended method is to increase the toughness and strength of the article and refine the case. The articles are allowed to cool slowly in the carbonizing pot to a temperature of about 650 degrees C. (1200 degrees F.), are then reheated to a temperature slightly exceeding the lower critical point of the case, which usually is from 775 to 825 degrees C. (1427 to 1517 degrees F.), and are then quenched in water or oil. They should be removed from the quenching bath before their temperature has fallen below loo degrees C. (212 degrees F.). By allowing them to cool slowly to a temperature of about 650 degrees C. (1200 degrees F.) and then reheating to a temperature of about 900 to 950 degrees C. (1652 to 1740 degrees F.), followed by quenching in oil, from which they are removed before they have dropped below a temperature of 100 degrees C. (212 degrees

F.), then reheating to about 800 degrees C. (1472 degrees F.) and again quenching in water or oil, both the case and the core will be thoroughly refined and their toughness greatly increased. In order to reduce the hardening stress created by quenching, the objects, as a final treatment, may be tempered by reheating them to a temperature not exceeding 200 degrees C. (392 degrees F.).

Results of Hardening without Reheating. — That part of the process where a most important improvement has been made in recent years is in the final hardening by quenching in a suitable bath. It formerly was customary at the end of the carbonizing period to open the pot and fling the contents headlong into a tank of cold water. Here and there some of the more careful workers took each article separately, but direct from the pot, and plunged it into water. These latter obtained better results, but even they had a great deal of trouble in the way of breakages and want of regular hardness. Finding that axles taken singly from the pot and quenched were better than those quenched in bulk, and that if allowed to cool down to cherry-red they were better still, an application of the old rule to harden on a rising heat led to the now established principle of allowing the pot and its contents to become quite cold, afterward reheating to cherry-red and quenching with water. By this means a case of great hardness with a very tough core is obtained — that is, of course, provided a suitable steel is employed.

To understand the reason of this improved method of working it must be remembered that the exterior of the steel is now of about 0.80 per cent carbon, and that steel of all kinds raised to and maintained at the high temperature employed for casehardening will, unless subjected to mechanical work, show evidence of overheating, being very brittle and liable to easy fracture; and though quenched m water, and consequently hardened, the metal has little or no cohesion and readily wears away. Steel so hardened breaks with a very coarse crystalline fracture, in which the limits of the case are badly defined. It is known that when steel is gradually heated there is a certain point at which a great molecular change takes place, and that perfect hardness can only be obtained by quenching at this critical point.

If quenching takes place below the critical temperature, the steel is not sufficiently hard; if above, though full hardness may be obtained, strength and tenacity are lost in part or completely, according as the critical temperature is exceeded by much or by little. This critical point lies between 1380 and 1470 degrees F., or cherry-red color heat. It may be asked why it is not sufficient, when taking the article out of the pot, to allow it to cool down to cherry-red and then quench it. To this the answer is that the high temperature has already created a coarsely crystalline condition in the steel, and that until it has become quite cold and has again been heated up to the critical temperature, a suitable molecular condition cannot be obtained. When steel is cooled suddenly, it bears in its structure a condition representative of the highest heat to which it was last subjected.

Casehardening in Cyanide. — The cyanide is melted in a cast-iron pot in a furnace and then the work to be casehardened is entirely immersed in the cyanide, which is heated to a bright cherry red. The work should be suspended by fine iron wires. Fifteen or twenty minutes after the work has been thoroughly heated through, it can be removed, and a casing of suitable depth for ordinary purposes is insured. The length of time of immersion will simply add to the depth of the casing, but thirty minutes of heating will give a very deep casing. The work can be dipped in clear cold water immediately after having been removed from the cyanide bath, or it may be permitted to cool, be reheated and hardened in the approved manner for casehardening. When small pieces are to be heated in cyanide, it is best to use wire baskets. These must be so made that the liquid has free access to all the surfaces of the finished pieces.

Local Hardening. — In many cases it is essential that the piece of work be hardened at a certain place and that other parts be left soft. There are three ways in which this can be accomplished: First, by copper-plating and enameling; second, by covering the part which is not to be hardened with fireclay; and third, by using a bushing or collar to cover the part to be left soft.

In the first case the article should be painted with enamel where it is to be hardened, the enamel being baked after having been applied. The remainder of the piece that is to be left soft is copper-plated. In the second case, if the article to be hardened has a recess, such as a hole, slot, etc., this may be filled with clay. The third method is used when a shaft, for example, is only to be left soft for a short distance. A collar is then placed on the shaft, and this provides the easiest and least expensive means for accomplishing the purpose.

In the case where enamel and copper-plating is used, the enamel will burn away and allow the surface covered by it to absorb carbon and, hence, to be hardened, whereas the copper will stand a very high heat and prevent hardening of those portions that are covered by it. If the copper is burned off, it is an indication that the work has been overheated. The clay prevents the hardening of a portion of the work in the same way as does the copper. It is also of advantage when dipping the work, as it prevents the formation of steam pockets which are apt to warp or distort the piece. When a sleeve or collar is used, this should be made about one-half inch longer than the part which is to be left soft, so as to prevent carbonization near the ends of the collar.

Casehardening Alloy Steels. — When nickel steels are heat-treated by casehardening, nickel seems to retard the process somewhat and the hardness of the "case" is somewhat lower than that obtainable in ordinary carbon steels. On the other hand, nickel tends to oppose the crystallization of the steel at high temperatures and to eliminate the consequent brittleness. With a 2 per cent nickel steel, the following temperatures are recommended: The steel should first be quenched from a temperature of 1830 degrees F. It is then given a second heating to 1380 degrees F., and is again quenched, after cooling to about 1290 degrees F. A single quenching from 1290 degrees F. gives the greatest hardness in the case, but not the greatest tenacity in the core. Quenching from 1380 degrees F. gives a somewhat higher tenacity but a slightly lower hardness in the case. A 6 per cent nickel steel should be quenched first from 1560 degrees F., and after reheating, from 1245 degrees F. Since this high nickel percentage almost completely prevents the brittleness of the core, one quenching from about 1290 degrees F. is, in most cases, sufficient. Steels with from 1 to 1.2 per cent chromium are sometimes used when an especially hard case is required. This element aids the crystallization of the core and the double quenching is necessary. Chrome-nickel steels with a low chromium content require about the same heat-treatment as pure nickel steels. A mixture of 60 parts wood charcoal and 40 parts of barium carbonate is recommended for carbonizing.

Casehardening Practice at the Juniata Shops of the Pennsylvania R.R. — The information in the following relates particularly to the method of packing the work, and to the use of test pieces to determine the quality and depth of the hardening. The following formula is used as a packing mixture: 11 pounds prussiate of potash, 30 pounds sal soda, 20 pounds coarse salt, 6 bushels powdered charcoal (hickory preferred).

The whole is mixed thoroughly, using about 30 quarts of water in the mixing; the above quantity is sufficient to harden three boxes of material containing the following parts: 2 links, 2 link blocks, 2 link-block pins, 2 valve-rod pins, 4 knuckle-joint pins, and 24 gibs for spring rigging.

The box required to hold these parts measures 40 inches long, 16 inches wide, and 12 inches deep; the time required to harden them properly is fourteen hours. Smaller parts, like link die plates, eccentric-rod jaw pins, and nuts below 1 inch, are usually packed in smaller boxes, or pipe 8 inches in diameter, which require between four and five hours heating. Link motion bushings and similar light parts are also packed in small boxes or pipe and require from two and a half to three hours.

Packing the Material. — The following method is pursued in packing the material: The bottom of the box is covered to a depth of 2 inches with the compound; the parts to be hardened are placed solidly, so that the compound is in contact with the bottom surface of the work; care is taken, however, that the work does not touch the sides of the box or other pieces in the box. After the first layer of the material is placed, it is covered on all sides and on the top with the compound, and is solidly packed with a suitable implement — a bolt with a large head will do. After the first layer is packed the same process is repeated, being careful to have sufficient compound between the two layers to prevent contact. There should not be less than 2 inches of packing material on top of the last layer. The lid which fits inside of the box is then thoroughly sealed with a luting of fireclay.

When in the furnace, the box rests on rollers to allow the flames to pass under it. The furnace is kept at a bright red heat, but not hot enough to scale or blister the work; when the material has soaked in the fire a sufficient length of time, the box is withdrawn to a trestle which is flush with the floor of the furnace, and stands parallel with and close to the water tank. The lid is then removed from the box and, if links are being hardened, the link is turned on edge and a bar passed through the slot in the link and lifted by two men, being plunged into the water endwise; the bar is then withdrawn and the link is allowed to remain in the tank until cold.

The results required in this case are obtained by emptying the contents of the box into the water at once. The tank used has a line of 1½-inch iron pipe connected with the service pipe, running around the four sides and close to the bottom, with ¼-inch holes drilled about 1½ inch apart. This supplies cold water to the work and drives the hot water to the top, where it is carried to the sewer by means of an overflow pipe.

The Use of Test Pieces. — With each box of material to be hardened a test piece is used. In the case of links, the test piece is 1¼ inch thick by 3 inches wide, and 12 inches long; this test piece is stenciled with figures giving the class of engine, the construction number and the date of hardening. After the links are taken out, the test piece is broken under hydraulic pressure and examined for depth of hardening, after which it is also subjected to a file test. Smaller parts are similarly tested. These test pieces are kept for two years or more for reference. It is possible to put from 1/16 inch to 5/32 inch depth of case on the link work in fourteen hours' time, and 1/16 inch on bushings and other small parts in from two and a half to three hours' time. All parts to be casehardened must be thoroughly cleaned.

Degree and Depth of Hardened Surface. — The percentage of carbon contained in the casehardened surface should vary according to requirements. A high-carbon case containing 1.1 per cent carbon gives a very hard wearing surface suitable for work that must withstand a fairly constant pressure, as shafts running in bearings, etc., but for parts which must withstand repeated shocks, this amount of carbon would render them too brittle, and in such cases it is advisable not to exceed 0.90 to 1 per cent carbon. For most purposes, 0.90 per cent carbon is preferable. Recent investigations indicate that the percentage of carbon in the hardened crust varies with the depth of the latter; the deeper the penetration, the higher the carbon content. Crusts about 0.050 inch deep usually have from 0.85 to 0.90 per cent carbon on the surface. In many instances, a penetration of 0.040 inch is sufficient, but if the work is to be ground after casehardening, it is advisable to carbonize to a depth of about 1/16 inch. Too deep a carbonized case makes the work more brittle, partly because of the prolonged exposure to a high temperature and partly on account of the increase in the hardened section and the decrease in the softer and more ductile core; hence, parts to withstand bending stresses, like gear teeth, should not be carbonized too deeply. The penetration of the carbon increases with the temperature and with the time of exposure, but not in direct proportion to these two factors. Carbonization takes place rapidly until the crust is saturated with carbon, when there is a sudden diminution in the rate of carbonization, which varies according to the temperature.

To Clean Work after Casehardening. — To clean work, especially if knurled, where dirt is likely to stick into crevices after casehardening, wash it in caustic soda (1 part soda to 10 parts water). In making this solution, the soda should be put into hot water gradually, and the mixture stirred until the soda is thoroughly dissolved. A still more effective method of cleaning is to dip the work into a mixture of 1 part sulphuric acid and 2 parts water. Leave the pieces in this mixture about three minutes; then wash them off immediately in a soda solution.

Straightening the Work after Hardening. — On account of the manner in which steel is rolled, drawn or forged, the density varies in different parts of the steel, and no matter whether the material is heat-treated or not, it will warp more or less when hardened. It is, therefore, necessary to provide apparatus for straightening the work. In straightening, it is necessary to bend the work about twice as much as would be required to merely keep it straight while the pressure is applied, as, on account of its elasticity, it will have a tendency to work back to its original form. Small rollers and shafts can best be straightened in a vise by having a three-point contact on the jaws. For large diameters a special straightener will be required. A surface plate placed to the height of a man's eye, and at a slight angle towards the light, provides the easiest means for testing work of this character while being straightened.

When there is a large quantity of rings to be straightened or trued up, a surface plate can be readily rigged up in the following manner: A solid strap is provided on one side and a compound lever on the other, adjustable to any place along the plate by means of a slot in the latter. By a slight movement of the lever the ring can be trued up. An indicator should be placed at the front of the plate so that the operator can try a ring to see at which points the ring is out, and also the amount necessary for making it round. In straightening washers or flat pieces of any kind, the hydraulic press provides the best possible means. It might be well to mention that washers or flat pieces should be ground by taking a small amount off each side alternately, as, otherwise, they will return to their original warped shape. Another precaution, relating to the grinding of cylindrical surfaces, is to use a copious supply of water, as otherwise the heat of the grinding operation will draw the surface, producing soft spots. These will appear to have been caused by improper casehardening, but as a matter of fact, they are wholly produced during the grinding operation.

Casehardening for Colors. — When hardening for colors, a furnace like the one shown in Figs. 3 to 6 is necessary. A very satisfactory method of coloring, which at the same time hardens deep enough for the class of work which is to be colored, such as wrenches, crank levers for automobiles, nuts, etc., is as follows: Mix 10 parts charred bone, 6 parts wood charcoal, 4 parts charred leather and 1 part powdered cyanide. The charred bone may be obtained by placing a few boxes of raw bone in the furnace on Saturday night (if the furnace is not run over Sunday). If much fire is in the firebox it should be drawn, as the heat in the furnace will be sufficient to char the bone to a dark brown. The charred leather may be obtained in the same way. The leather should be black, crisp and well pulverized, and the four ingredients should be well mixed together. The object in charring the bone and leather is to remove all grease. The parts to be colored must be well polished and they should not be handled with greasy hands. These rules must be observed if a nice class of work is desired.

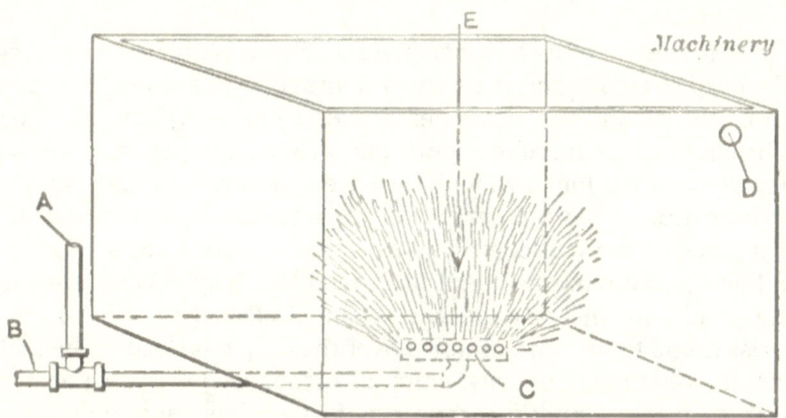

Fig. 14. Hardening Tank Arranged for Mottling and Coloring

If the colors obtained are too gaudy, the cyanide may be left out, and if there is still too much color, leave out the charcoal. When packing parts to be colored and hardened, they should be packed in a common gas pipe, for the reason that when dumping into the water the parts must not be exposed to the air, and a pipe is much easier to handle than any other shape. The open end can be brought down close to the top of the water before the parts are liable to come out, but not so with a box, for just as soon as a box is tipped a little the parts begin to fall out, and become exposed to the air.

In heating this class of work, heat to a dark cherry-red and keep at that heat for about four or five hours; if heated too hot, no colors will appear. To harden and color the work when dumped, a tank must be arranged as shown in Fig. 14. A compressed-air pipe *A* must be connected with the water pipe *B*, and a large cap *C* should be drilled full of ¼-inch holes on top and around the sides. An overflow *D* should also be supplied. Fill the tank with water, then turn on air enough to fill the tank with lively bubbles and dump the work in the center, at *C*. When the work is all dumped, pull up the sieve which should be in the tank for the work to fall into, pick the work out and place it in pails of boiling water drawn from the boiler; let it remain for five minutes and then remove it to a box of dry sawdust for half an hour; remove it from here and dust it off and give it, finally, a coating of oil.

Chapter Eleven - New Casehardening Methods

The Development of Casehardening by Carbonaceous Gas. — During the past decade there has been a great deal of investigation relating to the conditions under which the various carbonaceous gases may be used in place of the familiar solid carbonizing materials. The old well-known casehardening process described in the preceding chapter was the only one known for many centuries. It was used without a question of its superiority until the manufacture of armor plate became such a large industry that efforts were made to find a better, or cheaper, way of causing the carbon to penetrate this plate.

The first method employed was to place an armor plate in a pit and cover it with a layer of charcoal, and then lower another plate onto it. The cover was then put on the pit and the plates heated to (or baked in) a temperature that was sufficient to cause them to absorb the carbon from the charcoal. Gas was generally used for the fuel, owing to the ease of controlling the heat. The next method tried was to send a current of carbonaceous gas between the two plates, in place of the charcoal. This caused the carbon to "soak in" in less time and was found more economical. Later, electricity was used for heating the plates, and with the carbonaceous gas and electricity, the carbon penetration was found to be more uniform over the entire surface of the plate.

The results obtained from the action of carbonaceous gas on armor plate have been such that a muffle carbonizing furnace has been built and placed upon the market. This machine, illustrated and described in detail in succeeding pages holds the work in a revolving retort, through which is sent a current of carbonaceous gas. This retort serves as a muffle that is surrounded with the flames of the heating gases. With this furnace, small pieces can be carbonized in much less time than formerly, and at about one-half the cost as compared with packing in iron boxes and then baking in an oven furnace. All of the labor of packing materials is done away with; carbon will penetrate the metal in less time and more evenly; its depth and percentage can be controlled more easily; and the work can be heated to the carbonizing temperature more quickly and maintained there more easily. A steady flow of carbonaceous gas can be kept passing through the retort, and thus any depth of carbon can be obtained without repacking the work.

Comparison between Old and New Methods. — When considerable depth of carbon is required, this is impossible with the old method of packing with bone and charcoal in an iron box and sealing on the cover. This is due to the fact that only a certain amount of carbon is present, and the longer the work is baked, the more there will be in the steel and the less in the charcoal. When an equilibrium is established, no more carbon will penetrate the metal, and to obtain a greater depth, the work must be packed in fresh carbonaceous material and the heating repeated.

With the gas process, however, the percentage of carbon in the gas surrounding the work can be maintained at a permanent figure until the carbon has penetrated to the center of the metal, the percentage of carbon possible to impart to the steel being far above that which is used for any kind of commercial work. Some of the gases that have been experimented with are methane, ethylene, illuminating gas, carbon monoxide, carbon dioxide, and gases that are made from liquids like petroleum, naphtha and gasoline. Most of these gases have been used in combination with ammonia, in order to ascertain to what extent this would aid in the penetration of the carbon.

The Carbonaceous Gas. — From the numerous experiments that have been conducted, it has been found that carbon monoxide is far superior to any of the solid carbonaceous materials in the specific, direct carbonizing effect it has upon steel. It is also better than all other gaseous materials in this respect. Carbonizing materials that do not contain nitrogen cost only from one-tenth to one-twentieth of the nitrogeneous materials. It has been found, however, that nitrogen acts as a carrier for the carbon, and when it is not present, carbonaceous materials have a very weak carbonizing effect; some investigations have shown that the effect is absolutely nil without the intervention of gaseous carbon compounds. When solid carbonaceous materials are used, the specific effect of the nitrogen is very weak, and it is only when these contain a high percentage of the cyanogen compounds that they have any marked carbonizing effect.

While carbon monoxide is capable of rapid penetration, it has an oxidizing effect on steel, and is liable to form a scale that will spoil small work which cannot afterwards be ground. This oxidizing effect is more pronounced in chromium and manganese steels. When carbon monoxide alone is used for the carbonizing medium, there is a distinct demarcation between the carbonized zone and the core of the metal. This is also a detrimental feature, in that when the piece is hardened, it has a tendency to crack at this demarcation, causing the outer shell to peel off.

The Giolitti Process. — To overcome these bad effects of carbon monoxide, a new process has been developed by Dr. F. Giolitti, Genoa, Italy. In this process the work is packed with wood charcoal in a cylinder, and when heated to the carbonizing temperature, a current of carbon dioxide is injected into the cylinder. It was demonstrated that when a slow current of carbon dioxide traversed a mass of wood charcoal, the carbonizing gas was supplied with great rapidity and without any excess of carbon monoxide. Thus, an equilibrium with free carbon was established at the carbonizing temperature. The exhaust gas contained less than three per cent of carbon dioxide, it being almost entirely carbon monoxide, and its volume being about double that of the carbon dioxide which was introduced into the apparatus. Some results that were obtained with carbon monoxide alone, and in combination with charcoal, are shown in Table 1.

Depth from Surface at which Sample was Analyzed, Inches	Percentage of Carbon			
	Carbon Steel		Nickel-chromium Steel	
	CO Gas Alone	CO_2 and Charcoal	CO Gas Alone	CO_2 and Charcoal
1/64	0.70	1.17	0.86	1.16
3/32	0.67	0.81	0.87	0.81
3/16	0.53	0.34	0.75	0.50
5/16	0.39	....	0.59	....

Table 1. Results of Carbonizing Steel with Carbon Monoxide for Ten Hours at 2000 Degrees F.

With the use of this new process, a more rapid penetration can be obtained than with any of the solid or gaseous materials, except pure carbon monoxide. The carbon is evenly distributed in the carbonized zone, and the peeling of the outer shell, when hardened or tempered, is reduced to a minimum. Any desired depth of penetration can be obtained without renewing the carbonizing material, and there is absolute security against the introduction of any foreign substance. Variations in the percentage and depth of the carbon can be obtained by diluting the carbon monoxide in nitrogen; by limiting the contact of the solids with the metal; and by varying the temperature during the carbonizing operations. Expansion or contraction and warping of the pieces being carbonized has also been reduced to a minimum. In fact, there are so many features that make it superior to the old method of packing and sealing the work to be carbonized in iron boxes, that it is safe to say that the new process is incomparably better.

After much experimenting, Dr. Giolitti decided that the double muffle furnace, shown in Fig. 1, was the most efficient and economical for carbonizing small or medium size pieces of varying shapes. By providing two muffles, one could be filled with work and kept at the carbonizing temperature, while the other was being emptied and refilled. If the amount of work would warrant, many more muffles could be used in the same furnace. The muffle to the right is shown by a sectional view through the center of the furnace, but the retort that holds the work is not sectioned, while the muffle to the left is shown by a sectional view on the center fine, thus revealing the interior.

Description of the Giolitti Furnace. — Cylindrical muffles A are made of some refractory material and are built into the brick-work of the furnace. Surrounding them are passages *B*, in which the combustion of the heating gases takes place. These passages are lined with firebrick or fireclay and the furnace is provided with regenerators, so that the fuel gas, which is furnished by producers, can be used in the most economical manner. By this arrangement and the valves that are supplied, it is possible to maintain the work at a uniform predetermined temperature. The work-holding retort *C* is made of seamless steel tubing and sets into flange *D*, which latter is attached to a frame

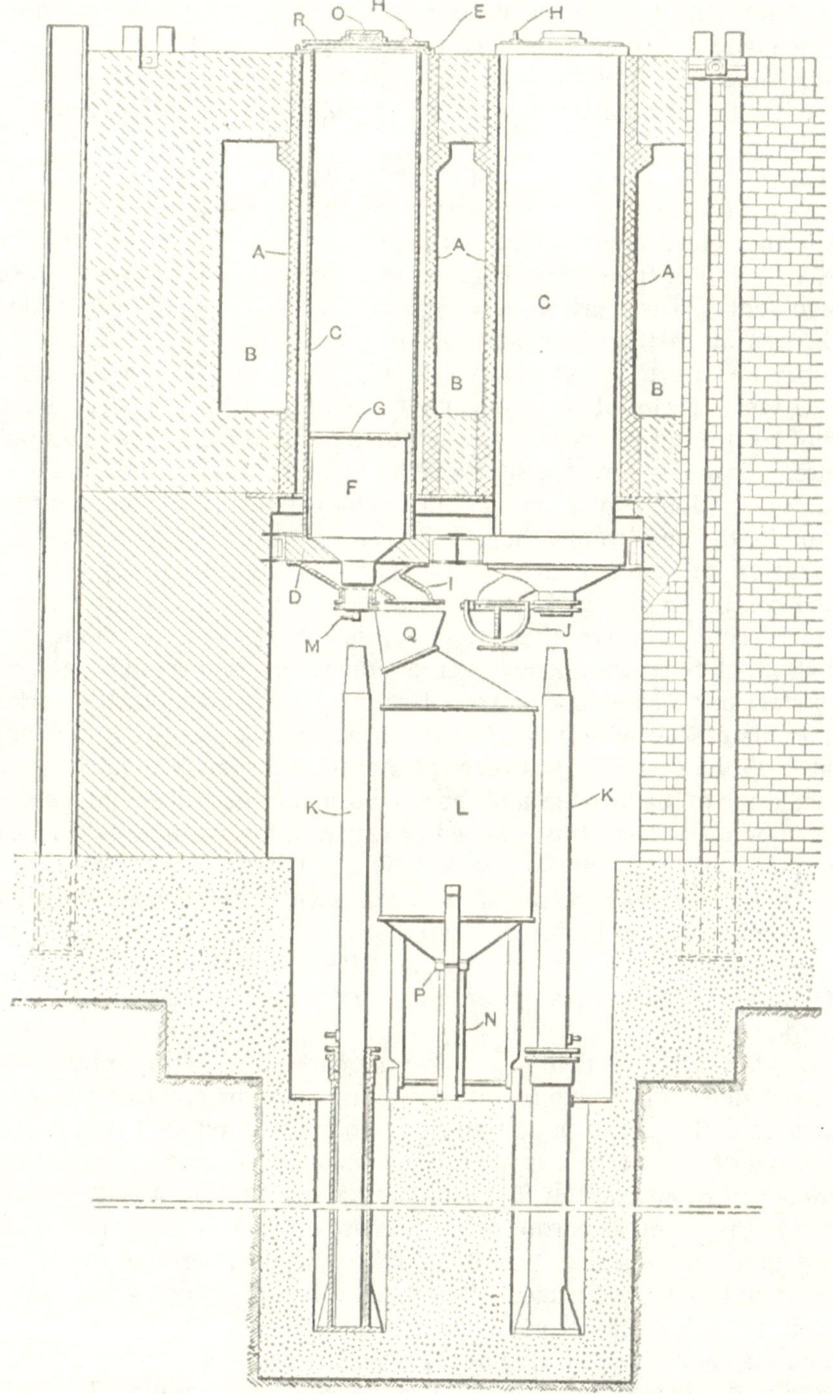

Fig. 1. Muffle Furnace for Carbonizing Steel with Charcoal and Carbonaceous Gas

that fits into the brick-work underneath the heating chamber. Flange *E*, which is made U-shaped to hold cover *R*, supports retort *C* at the top of the furnace. Thus, it is only the work of a few minutes to take out retort *C* and replace it with a new one, when it has become warped out of shape or burnt through.

Inside of retort *C* is located a hollow cast-steel cylinder *F*, and inside of this is located a device that evenly distributes the carbonizing gas around the work in the retort, by sending it through cover plate *G*, which is filled with holes. When all of the carbon is taken from the gas, it is allowed to escape through vent *H*. The work to be carbonized is stacked up on cover plate *G*, which rests on casting *F*, and this, in turn, rests on the same flange *D* that supports the retort. To the bottom of this flange is attached the cast-iron nozzle *I*, which is closed at the bottom by the non-return valve *J*. Underneath the muffles are located two hydraulic rams *K* and a cylindrical iron tank *L*, mounted on wheels, for handling the solid carbonizing material, which is in a granular condition. Tank *L* may be turned on its wheels so that spout *Q* will come under the nozzle *I* of either muffle.

Operation of the Furnace. — In operating this furnace, a continuous method can be employed. When a batch of work has been carbonized, the granular carbon is drawn off through nozzle *I* into tank *L*. After that, pipe *M*, through which the carbonizing gas enters the retort, is unscrewed, and ram *K* raises steel pot *F* towards the top of retort *C*, thus pushing the work that rests on plate *G* up with it, so that it can be removed as the ram proceeds upwards. When casting *F* has reached the top of its stroke, which is a little below the top of the furnace, and the old work has been taken out, new work is placed on disk *G*, which is lowered by ram *K* as fast as the work is located. When the retort is filled with work, tank *L* is wheeled out from under the muffles and raised over the top of the furnace with a hoist on a swinging arm. When over the top of the furnace, pipe *N* at the bottom of tank *L* is lowered into opening *O* in the center of cover *R*. Valve *P* is then opened to allow the hot granular carbon to flow out of tank *L* and fill the interstices surrounding the work.

If granular carbon is held at or near the carbonizing temperature, it acts very much like a liquid, and readily flows into all of the crevices surrounding the articles in the retort that are to be casehardened. When it is drawn off at the bottom of the retort it is at the carbonizing temperature, and the time consumed in removing the finished work and replacing it with new is so short that the granular carbon does not cool down to a temperature below 1500 degrees F. Thus, it retains its mobility and flows around the work. An operator on top of the furnace might assist this flow by using iron rods that can be inserted into the retort through holes in the cover.

When the retort is properly filled, butterfly valve *P* is closed and tank *L* is lowered and wheeled to its position underneath the muffler. The work is then allowed to stand until the carbonizing temperature has been reached. In the meantime, pipe *M* has been screwed into position, and when the carbonizing temperature has been reached, the carbonaceous gas is injected into the retort through this pipe. While the work in the retort to the left is being carbonized, the retort to the right can be emptied and filled, without in any way disturbing the process in the other.

Effect of Compressing the Gas. — Another method that has been tested by the designers of this furnace is that of compressing the carbonizing gases, and some very good results have been obtained. The tests demonstrated that when carbon monoxide acts on ordinary steel in the presence of free carbon, as in the furnace shown in Fig. 1, an increase in the depth of carbonization will be obtained with an increase of the pressure on the gas, and there will also be an increased concentration of carbon within the carbonized zone.

In carrying out some experiments of this nature, a cylindrical retort was used, like that shown in the sectional view in Fig. 2. In this, electricity was used to heat the work to the carbonizing temperature. The work was packed in charcoal in a retort into which a current of carbon dioxide was injected, in a very similar manner to the method used in the furnace in Fig. 1. In the illustration, *A* and *B* are the clamps for the terminals and these conduct the current to nickel-wire spiral *D*. This wire is wound around porcelain tube *E*, which can easily be inserted into, or taken out of, the apparatus. By taking off nut *N*, tube *E* is readily slipped into fireclay tube *F*, which is surrounded by steel tube *G* and in-

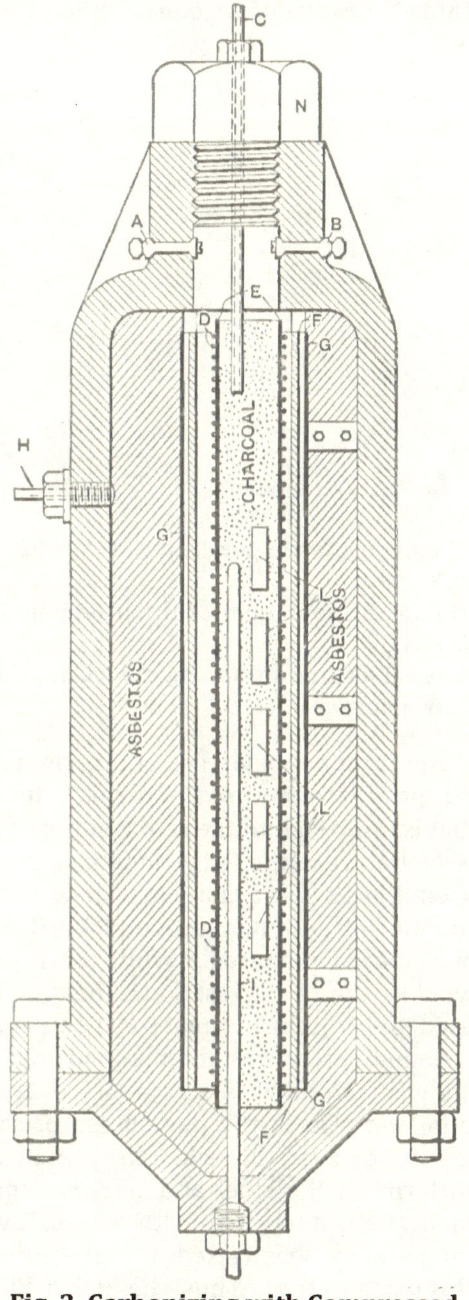

Fig. 2. Carbonizing with Compressed Carbonaceous Gas, using Electric Heating Means

sulated with asbestos.

Table 2. Results of Carbonizing Steel with Compressed Gas for Three Hours

Kind of Steel	Carbonizing Temperature, Degrees F.	Pounds Pressure of Carbonizing gas	Percentage of Carbon at a Depth of	
			1/64 Inch	1/32 Inch
Nickel steel*	1600	235	0.71	0.12
	1775	235	0.99	0.29
	1650	400	0.73	0.36
Chromium steel†	1750	235	2.22	1.03
	1925	235	3.10	1.39
	1875	400	2.37	1.40
Chrome-nickel steel‡	1525	235	0.45	0.54
	1650	235	0.76	0.49
	1525	400	0.54	0.56

[*] Composition in per cent: Nickel, 5.02; carbon, 0.118; silicon, 0.20; manganese, 1.53.

[†] Composition in per cent: Chromium, 2.33; carbon, 0.41; silicon, 0.15; manganese, 1.02.

[‡] Composition in per cent: Nickel, 3.17; chromium, 1.5; carbon, 0.33; silicon, 0.06; manganese, 1.15.

The carbon dioxide gas enters the retort through tube *C* and the used gas escapes through pipe *H*. Porcelain tube *I* contains a thermo-electric couple that is inserted into the retort at the opposite end from gas tube *C*. With it the temperature of the entire length of the casehardening chamber can be measured. Blocks *L* are the experimental pieces to be carbonized, and are surrounded by granular carbon. Table 2 shows some results obtained with various kinds of alloy steels. While this is only a crude experimental apparatus, it would seem to suggest some ideas or principles that can very profitably be used for carbonizing steel parts on a commercial scale.

Definite proof was obtained that variations in the pressure of the carbonizing gas were always accompanied by variations in the depth of carbonization, and also in the percentage of carbon in the carbonized zone. It was found, however, that when the pressure was too high it would cause an oxide to form on the steel and this was more pronounced with chromium and manganese steels than with others. It was also found that as the carbonizing temperature was raised, the pressure could be increased without causing this oxide to form. Thus, the higher the carbonizing temperature, the higher can be the pressure used on the carbonizing gas, with an absolute assurance that no oxidation will take place.

A rod of soft steel, 2 3/8 inches in length and three-eighths inch in diameter, was casehardened for about three hours by heating three-fourths of an inch of its central portion to about 1800 degrees F. and allowing this temperature to decrease towards the two ends, so that at these the temperature was about 900 degrees F. Unmistakable carbonizing took place in all portions that were above 1450 degrees F. The surface was absolutely unaltered in the hottest portion in the center, which was also the most intensely carbonized, while a distinct layer of oxide was seen in the cooler portions, this oxide thickening as the temperature lowered towards the ends.

Such good results were obtained by compressing the carbonizing gas as it was injected into the bed of charcoal in which the work was packed in the retort, that this method promises to become a commercial success. While the mixed agent, carbon monoxide and charcoal, increased both the speed of penetration and the percentage of carbon in the carbonized zone over all previous methods or materials used for carbonizing steels, compressing the carbon monoxide has still further increased these factors. Like all methods and processes, however, it must be handled properly. The amount of compression, as well as the carbonizing temperature, varies with different kinds of steels. Therefore, these must be discovered and properly adjusted, if work is to be turned out that is free from oxide and scale, and that has the desired penetration uniformly distributed over all portions of the exposed surfaces.

Whether work is carbonized with an ordinary flow of carbon dioxide into and through the charcoal, or by compression, the advantages which vertical muffles, as shown in Fig. 1, have over horizontal muffles are in the greater speed of charging and removing the work, due to the greater simplicity of the operations, the uniformity of the treatment of all pieces forming the charge, and the more uniform distribution of the carbonizing gases, due to the spaces being reduced to a minimum.

Time Required for Operation. — The time for the various operations with the furnace shown in Fig. 1 is as follows: Charging the pieces to be carbonized, from 1 to 5 minutes, according to their size and shape; completely filling the retort with granular carbon, from 1½ to 4 minutes; lowering ram *K*, replacing pipe *M*, removing tank *L*, and closing down cover *R*, 1 minute; drawing the granular carbon from retort *C* into tank *L*, 4 minutes; raising ram *K* and removing the work from the retort, 2 minutes. The time consumed in all the operations, where ordinary work is being handled would, therefore, be about ten minutes, but with specially shaped pieces and unfavorable conditions, this time might be extended to 30 minutes.

By pre-heating the work to a carbonizing temperature before putting it into the retort, no time will be lost in fully heating it in this furnace. The temperature of the granular carbon can then be maintained nearly up to the carbonizing temperature, as it will not be chilled by cold work. The process can thus be made strictly continuous. Under these conditions, a depth of carbon penetration of 1/32 inch can be given the work in one hour, and of 1/16 inch in two hours. Thus it will be seen that from 1¼ to 2½ hours is all that is required for the complete carbonizing operations in one retort.

By using gas for fuel and gas for carbonizing, the work can be controlled within closer limits than with any other process, unless it should be an electric one, and the arrangements of this furnace are such that its capacity for producing work is greater than that of any furnace which has been designed with the same size of work holder. If there is not enough work to keep both muffles going on pre-heated work, one of the muffles can be used as a preheating furnace, while the other is doing the carbonizing. By alternately using the muffles for preheating and carbonizing, an amount of work will be turned out that will compare favorably with any other casehardening furnace. If desired, the current of carbonaceous gas can be used for a whole or any given part of the carbonizing time, and thus the results obtained can be made to cover a wide range. Where localized casehardening is required, the granular carbon can be drawn off until only enough is left to insure the chemical equilibrium in the gas, and by thus isolating the carbon monoxide, it will intensify its specific action.

Apparatus for Casehardening by Gas, Suitable for Small and Medium Work. — The casehardening process described on the preceding pages is intended for specially large work. The American Gas Furnace Co. has brought out a casehardening plant, using gas as the carbonizing material, which is suitable for small and medium work. Briefly stated, the process as performed by the apparatus brought out by this company consists in placing the work in a slowly revolving, properly heated, cylindrical retort into which the carbonizing gas is injected under pressure. From the gas, the work absorbs the volatile carbon. The absorption of carbon begins as soon as the work is sufficiently heated to attract it, and continues throughout the process, because the work is constantly and uniformly exposed to a carbon charged atmosphere under pressure, instead of to solid carbonaceous material which turns to ashes wherever it is in proximity to the heated parts. All the parts of the same piece of work and all the pieces contained in one charge in the retort are, therefore, continuously subjected to exactly the same condition as regards the presence of carbon, and the result is a uniformity and speed of operation not obtainable by other methods.

The complete gas casehardening plant consists of a generator of carbonizing gas and revolving retorts used for the carbonizing process. The generators are generally made in sizes to supply two or more machines with carbonizing gas. The gas is produced from refined petroleum and the carbon vapor is so diluted by a neutral gas that the proportion of carbon that is supplied to the work is not greater than that which can be absorbed by the work without forming obstructive carbonaceous deposits. The carbonizing machine proper consists of a carbonizing retort and a cylindrical furnace body, in which the retort is enclosed and in which it rotates. Suitable arrangement is made for charging and discharging the work. The furnace for the exterior

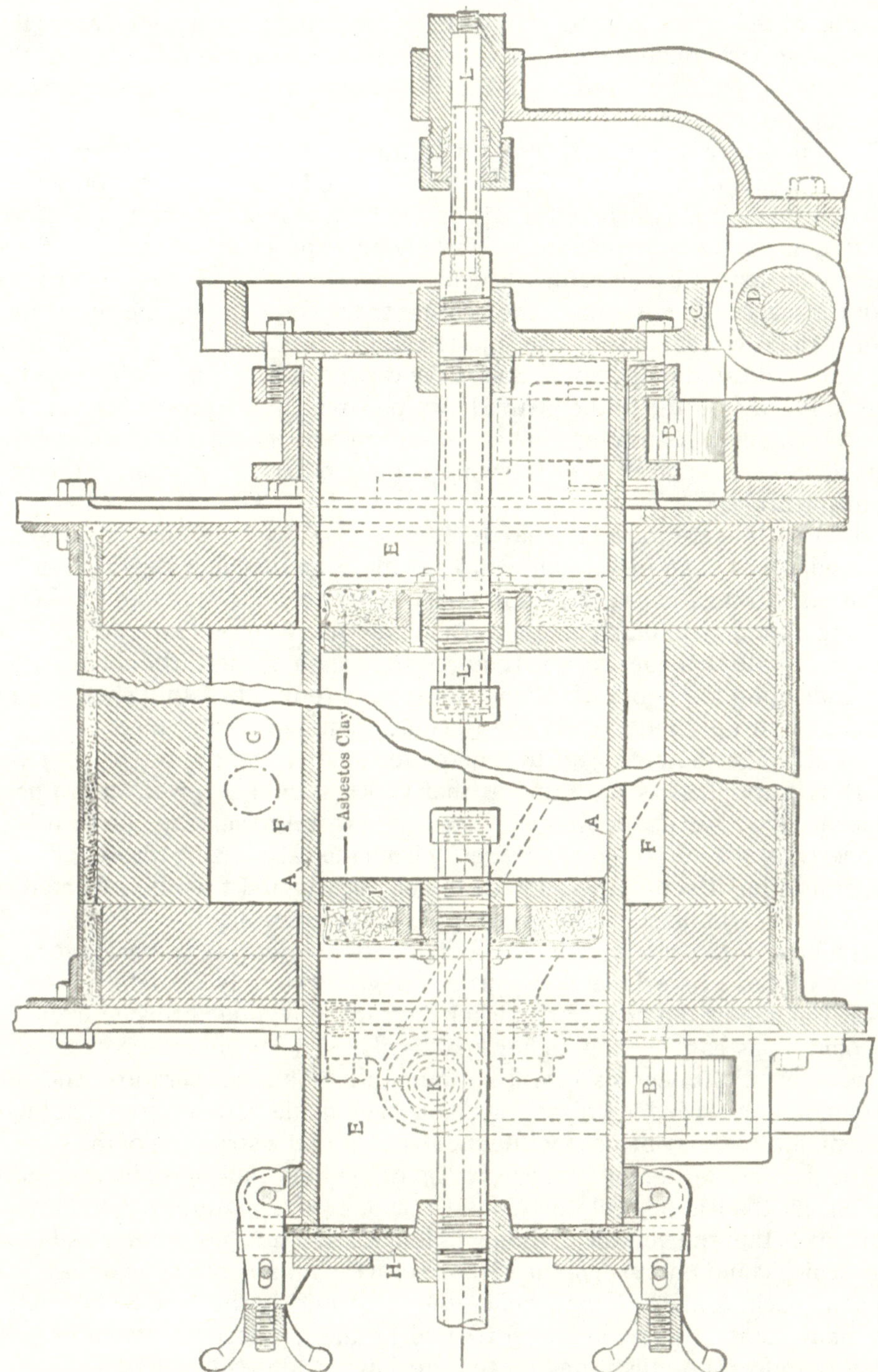

Fig. 3. Receiving cylindrical retort brought by the American Gas Furnace Company for casehardening by gas.

heating of the retort is fired with fuel gas requiring a positive air blast. Besides these carbonizing retorts, of course, ordinary furnaces for reheating the work for hardening are required. This reheating can also be done in the carbonizing retort, if desired.

The line-engraving, Fig. 3, shows a sectional view of the carbonizing machine. A wrought-iron retort is shown at A which is slowly rotated on rollers *B* by worm-gear *C*, which, in turn, is driven by worm *D*, the shaft of which is rotated in any suitable manner, preferably by a sprocket and chain. At *E* are shown air spaces in the retort formed by two pistons *I* between which the work is placed. At *F* is shown the heating space surrounding the retort into which the fuel gas and air are injected under pressure from two rows of burners, indicated in the upper half of the casting at *G*. The cover *H* closes the retort. It is connected to the piston *I* by pipe *J*, which also provides a vent for the retort. Cover *H* and this pipe are withdrawn to charge and discharge the retort, and are replaced after the work is inserted, before beginning the carbonizing process.

Steel to Use for Gas Casehardening. — For casehardening by the gas method, it has been found that articles made from machine steel containing from 0.12 to 0.15 per cent carbon give the best results, although steel containing from 0.20 to 0.22 per cent carbon may also be used to advantage. The length of time that the work is required to remain in the carbonizing retort depends upon the depth of carbonized surface required. A thin shell will be produced in one hour, while the thickness will constantly increase if the work is left in the retort up to nine or ten hours. The treatment after the work is carbonized is the same as that which should be given to ordinary casehardened work. As already stated, it is rarely the case that work is properly hardened, if quenched directly from the carbonizing retort, but, as a general thing, it should be allowed to cool slowly and then be reheated to harden the carbonized surface at the proper hardening heat.

The heat of the retort while carbonizing the work must be varied for different classes of steels, and the proper degree can only be determined by trial. The higher the heat, the quicker the carbon will be absorbed from the carbonaceous gas, but the higher heat tends to make the core coarse. As a rule, about 1500 degrees F. will be found a suitable temperature, and this should not be exceeded unless tests have been made to determine that higher temperature may be used without detriment to the structure of the steel.

The gas casehardening process can be carried out more rapidly and more uniformly than is possible with solid carbonaceous materials. Another advantage is that the volatile carbon will find its way into slots, holes and cavities which could not receive the carbon from the granulated bone or any other solid packing material, and, hence, the uniformity of the product is greater. In many cases, low-carbon steel treated by the gas casehardening process may, therefore, be substituted for tool steel in machine construction.

Chapter Twelve - Heat-Treatment of Gears for Machine Tools

In the earlier days of machine tool construction, when carbon tool steel was used for cutting, and relatively light work was the order of the day, cast-iron gears were used for transmitting power. With the advent of air-hardening tool steel and heavier work, the use of mild steel gears became necessary, while to-day, with tools of modern high-speed steel the use of heat-treated alloy steel gears is well nigh imperative. Gears of this last class may be divided into two general groups — casehardened gears, with a low-carbon soft center or core and a high-carbon hard exterior or case, and hardened high-carbon gears which are of the same composition and hardness throughout. The characteristics, heat-treatments and merits of these two groups, as viewed in the light of a wide experience with gears used in motor-car construction, will be discussed briefly in the following.

Casehardened Gears. — Case-carbonized gears may be made from four general classes of steel, *viz.*, straight-carbon, nickel, chrome-vanadium and chrome-nickel steel, and of each of these classes several modifications will be found in the market. On the whole, the steels containing chromium are to be preferred, for they are freer from the tendency to lamination shown in nickel-steel (especially 3½ per cent nickel steel) and they also absorb carbon more easily, thereby lessening the length of time and expense of carbonizing. Before carbonizing, the carbon content of each of the steels mentioned, should be about 0.20 per cent, and never more than 0.25 per cent, to avoid brittleness in the teeth. The carbon in the case should be raised to about 0.90 per cent, which can readily be done by the proper selection of carbonizing material and by using the proper temperature for carbonizing.

The temperature for carbonizing, in general, should be about 1600 to 1650 degrees F. for all the classes of steel previously referred to. Lower temperatures do not give sufficient depth of "case," unless the heating operation is much prolonged. On the other hand, higher temperatures result in a case of excessive carbon content and in a core of such large-grained size that it will not respond to the subsequent heat-treatment as readily as if a temperature of 1600 to 1650 degrees had been used.

The heat-treatment after case-carbonizing is the most important part of the process, and upon it depend the physical properties of the finished work. As already stated, after carbonizing we have a piece of steel with a 0.20 per cent carbon core, and a 0.90 per cent carbon case, and the object of the treatment is to put both the core and the case into the best possible physical condition. Both need refining to correct the large-grained structure developed by subjecting the steel for many hours to the carbonizing temperature. Since the refining or hardening temperature of the core is about 200 degrees F. above that of the case, this difference determines the most approved method of heat-treatment.

The old method consisted in quenching the piece in oil or water at the end of the carbonizing operation, right from the box, at the temperature used for carbonizing. This resulted in a large-grained core that was neither strong nor tough, and an overheated granular case which was hard, but which would not stand up in service any better than an overheated piece of tool steel. The first improvement on this old method was to allow the piece to cool in the box after it was removed from the carbonizing furnace, and then to reheat it to the proper temperature for hardening the case and quench in a suitable fluid. This procedure, however, did not develop a strong tough core.

The proper heat-treatment for casehardened gears is the so-called double treatment by which the pieces are first allowed to cool in the box after carbonizing, next reheated to from 1550 to 1625 degrees F. and quenched in a suitable medium to refine the core, then reheated to from 1350 to 1425 degrees F. and again quenched in a suitable medium to harden the case, and finally drawn in oil at not above 400 degrees F. to further increase the strength and toughness of the casehardened gear. The temperatures given are approximate only; for exact information concerning any particular steel, the user should consult the steel-maker.

Fig. 1. Cored Pot or Box for Casehardening

There are many case-carbonizing compounds on the market and most of them have some merit. Those of bone are probably the least desirable owing to their lack of uniformity which results in uneven carbonizing. The most desirable are those consisting of definite mixtures of carbon and carbonates; they carbonize uniformly, and most of them can be used repeatedly without losing their power of giving up carbon to the metal.

The wear and tear on carbonizing furnaces, the fuel consumed, and the expense of the boxes are three important items in the cost of casehardening. In many cases it is possible to reduce all these items by the use of a cored instead of a solid box, as shown in Fig. 1. The proportion, of course, will vary with the work to be done, but if the general idea is worked out for each specific instance, it will be found not only that the cost of carbonizing is diminished, but also that the carbonizing is more uniform.

Hardened High-carbon Gears. — Unlike casehardened gears, hardened high-carbon gears are of uniform carbon content throughout, and, when hardened, have a uniform hardness throughout the tooth-section. The steels used for these gears are of three general classes, viz., silico-manganese, chrome-vanadium and chome-nickel steel — the last-named, in its several modifications, being by far the most used. The carbon content varies for the different classes from 0.40 per cent to 0.60 per cent. The heat-treatment of all these steels is very simple, consisting merely in heating the gear slowly and uniformly to the hardening temperature, which is usually about 1500 degrees F., quenching in oil, and afterward drawing in an oil bath. The result is strong, tough, dense-grained steel gears; these have been used with marked success in motor-car work, and are fast replacing soft steel and casehardened gears in machine-tool construction. Viewed from the standpoint of physical properties in the finished gear, the evolution in gear material from cast iron to tempered steel, may be seen in the following figures:

Material	Elastic Limit, Pounds per Sq. In.	Hardness — Scleroscope	Toughness — Guillery Impact Kilogrammeter
Cast iron	20,000	25	Negligible
Soft steel	40,000	35	2
Casehardened steel; (average test of alloy steel)	120,000	85	2.5
Tempered steel; (average test of alloy steel)	225,000	75	5

Comparison of Results. — For machine tools, hardened high-carbon alloy-steel gears appear to be preferable to casehardened gears for a number of reasons:

1. Physically they are stronger and tougher and should therefore be better able to resist sudden impacts and extraordinary loads. They do not show by file and scleroscope test the same degree of hardness as casehardened gears, but, nevertheless, with proper design, the dense-grained gear-tooth resists wear most satisfactorily, as was demonstrated recently by the examination of a motor-car transmission that had covered over 100,000 miles. The high-carbon steel gears in this car still showed the original tool-marks. Not long ago a designer of machine tools commented on the apparent softness of some hardened high-carbon gears, but found after several months of hard service that they still showed tool-marks, thus proving hardness ample for wear.

2. In service, especially for "clash gears," the superiority of these gears is most marked. On the clashing faces, casehardened gears are likely to have the hard case chipped off, thereby exposing the soft core to the impact of clashing. The hard chips fall into the gearing and may find their way into bearings, thus causing trouble. High-carbon steel gears with a uniform hardness throughout do not chip, nor do they "dub over."

3. The heat-treatment of high-carbon steel gears is much simpler than that required for proper casehardening. It is shorter, less costly and produces a more uniform product, and as the gear is heated but once for hardening, as

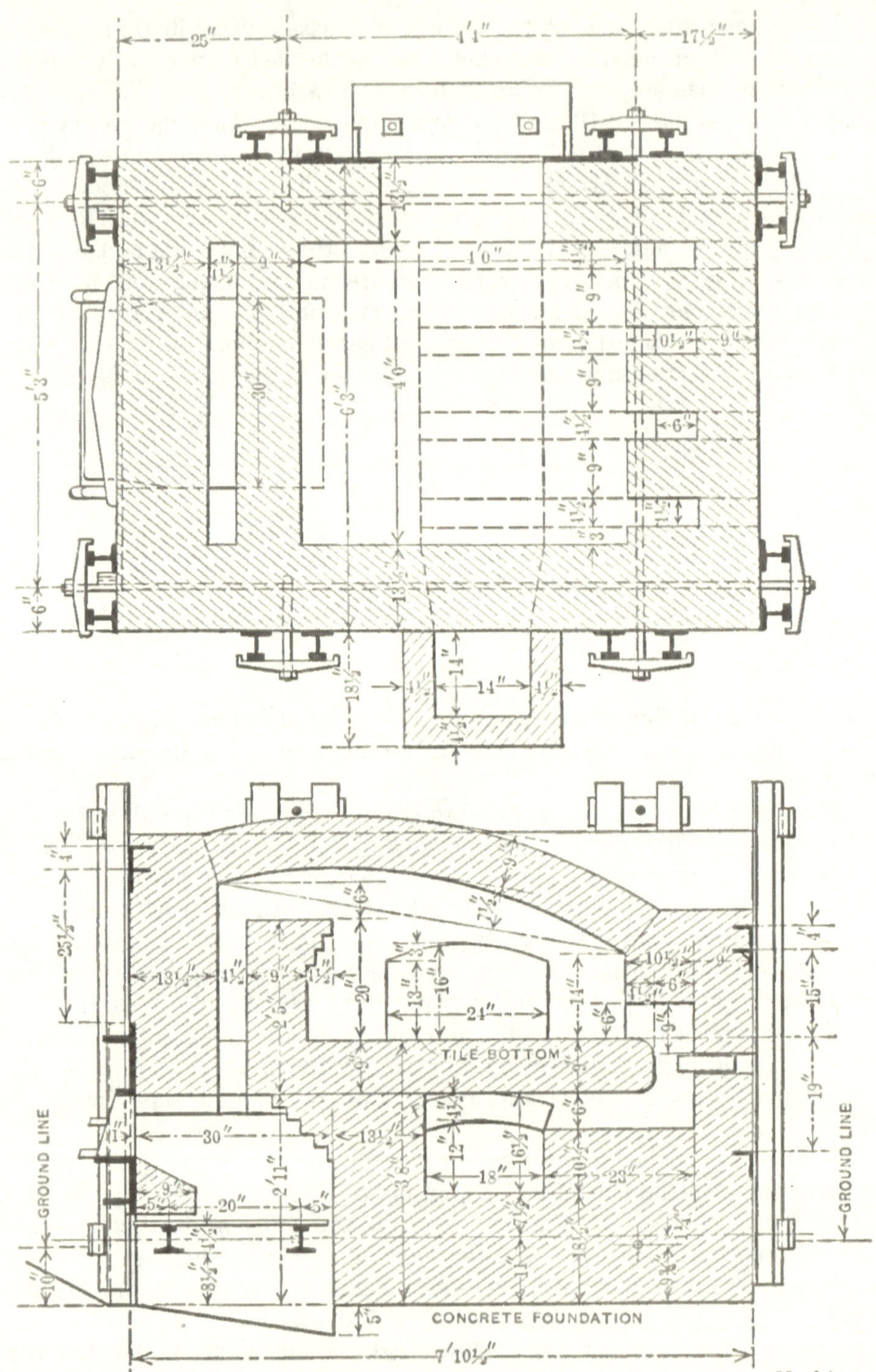

Fig. 2. Coal or Coke Furnace for Heat-treating Gears

compared with three times for casehardening, the finished gear is certain to be freer from warpage. The cost of proper casehardening is not generally appreciated, but it has been found that a casehardening steel must cost three to four cents per pound less than a regular high-carbon hardening steel, if finished gears made from both materials are to cost the same.

With all heat-treated gears, little points in design are important. The gear-teeth should not be undercut, for if the section at the root-line is smaller than at the pitch-line, greater hardness and brittleness is produced where least desired. Great differences in section should be avoided wherever possible, so as to do away with excessive warpage. Sharp edges and angles, even in key-ways, are the cause of internal hardening strains which frequently result in failures; hence, wherever possible, a fillet should be used in place of a sharp angle.

Furnace for Heat-treating Gears. — When heat-treated gears are suggested to the machine-tool builder as a remedy for some of his troubles, and as a means of eliminating an excessive item for replacements and repairs, one of his first questions naturally is, of what does a heat-treating equipment consist? Usually the second question is, what will it cost? The first item in equipment is a furnace. There are offered for sale a number of types of gasand oil-fired furnaces, but few are located in natural gas districts, and the price of fuel-oil has almost driven the oil-fired furnace from economic use. Coal and coke, however, are available everywhere, and a furnace using fuel of this kind is shown in Fig. 2. Anthracite, bituminous coal, and coke work equally well. When bituminous coal is used, the consumption with ordinary firing should not exceed 500 pounds in a twenty-four hour day. The cost depends somewhat upon the price of labor, but should not exceed $300.

Quenching Baths. — The next item is a proper quenching medium. Running water with a suitable tank is always necessary in a hardening room. To take the chill from the water in winter, or to raise the temperature a little at any time, a jet of live steam in the incoming water pipe will be found very convenient. The heat-treatment of alloy gear-steels requires an oil bath, the size of which depends entirely upon the amount of work to be quenched and the facilities for keeping the oil cool. The kind of oil best suited for oil-hardening was the subject of an investigation conducted by the laboratory of the Carpenter Steel Co., with the following results, comparison being made with water as a standard:

Hardening Medium & Numeric Hardening Quality

Water - 1.000
Mineral No. 1 - 0.2409
Mineral No. 10 - 0.2304
Corn - 0.1927
Mineral No. 2 - 0.1607
Cotton-seed - 0.1606
Fish - 0.1490
Rosin - 0.1350

For hardening, several of the mineral oils are more effective than fish and cotton-seed oils, which for a long time were looked upon as the best oils for this purpose. Mineral oil No. 1 has a specific gravity of 0.860, a flash point of 420 degrees F., and a viscosity of 170 seconds at 100 degrees F., as shown by the Saybolt viscosimeter. A mineral oil to this specification can be bought very cheaply.

Oil can be cooled by blowing cold air through it, or by pumping the oil through a coil of pipe immersed in cold running water, thus maintaining a circulatory system which admits cool oil at the bottom of the hardening tank and pumps the warm oil from the top through cooling coils back to the bottom. When air is used, care should be taken to avoid an excess during the quenching of a piece, for if air instead of oil were to strike the piece constantly, uneven hardening might result.

Drawing or Tempering Bath. — The next equipment item is a drawing bath. This may consist of oil, lead, or a combination of salts, contained in a cast-iron or steel vessel. The container is usually of very simple design and may be fired by gas, oil or coal. The oil should be a mineral oil with a flash point of not less than 600 degrees F., this temperature usually being sufficient for all temper-drawing purposes. If higher temperatures are desired, a mixture of two parts potassium nitrate and three parts sodium nitrate may be used. This mixture melts at 450 degrees F. and may be used for temperatures up to 1000 degrees F., or lead, which melts at 630 degrees, may be substituted. To indicate the temperature of the bath, a mercury thermometer should be used rather than a pyrometer, for most pyrometers will show considerable error at drawing temperatures under 800 degrees F.

Application and Calibration of Pyrometers. — The last item is a pyrometer. There are a number of good thermo-electric pyrometers on the market, and more depends upon the care of the instrument than upon the selection of any particular make. Following are a few rules for the use of the pyrometer and a simple method of calibration:

1. Keep the hot end of the thermo-couple as near the work as possible; do not put it through the furnace wall or roof, exposing the end to the direct heat of the flame, but place it so that it will attain, as nearly as possible, the same temperature as the work.

2. Keep the cold end of the thermo-couple protected from the direct or radiating heat of the furnace; that is, keep it cool.

3. Protect the voltmeter by a dust-proof case, and place it on a support free from vibration.

4. All switches should be of the wiping-knife type. Improper contact at the switches is a prolific source of error, and such errors are not readily located.

5. Carefully check all thermo-couples as soon as they are received from the manufacturer and before putting them into service. Adhere closely to this rule, instead of assuming that new thermo-couples are sure to be correct. New thermo-couples should not be used on faith without checking, since

they occasionally show a considerable error, and any one making use of them as standards will sooner or later come to grief.

6. Carefully standardize each pyrometer at definitely stated intervals — at least once a week, and as much oftener as possible. Frequent calibration is a matter not of convenience, but of necessity.

How can the calibration of a pyrometer be accomplished readily and accurately without the use of an extensive laboratory equipment? To this question of immediate interest in every hardening-room, the answer is that the easiest and most convenient method is based upon determining the melting point of common table salt (sodium chloride). Chemically pure salt, which is neither expensive nor difficult to procure, should be used where accuracy is desired. The salt is melted in a *clean* crucible of fireclay, iron or nickel, either in a furnace or over a forge fire, and is then further heated until a temperature of about 875 degrees to 900 degrees C. (1607 to 1652 degrees F.) is attained. It is essential that this crucible be clean, because a slight admixture of a foreign substance might noticeably lower or raise the melting point. The thermo-couple to be calibrated is then removed from its protecting tube and its hot end is immersed in the salt bath. When this end has reached the temperature of the bath, the crucible is removed from the source of heat and allowed to cool, and cooling readings are taken every ten seconds on the voltmeter. A curve is then plotted by using time and temperature as coordinates, and the temperature of the melting point of salt, as indicated by this particular thermocouple, is noted at the point where the temperature of the bath remains temporarily constant while the salt is freezing. The length of time during which the temperature is stationary depends on the size of the bath and the rate of cooling, and is not a factor in the calibration. The true melting point of salt is 801 degrees C. (1474 degrees F.), and the needed correction for the instrument under observation can be readily applied.

Cost of Equipment. — The cost of this equipment, including a coal-fired furnace, as shown in Fig. 2, five to seven barrels of hardening oil, one barrel of drawing oil, the tanks for holding these oils and a pyrometer, should be about $500 to $600. The equipment just noted is the one necessary for tempered gears. When casehardened gears are heat-treated, there is necessary, in addition to this, case-carbonizing boxes and a carbonizing compound. A second furnace may also be necessary, depending upon the quantity of casehardened gears to be treated. It is thus seen that for casehardened gears, the heat-treating equipment is more expensive than that required for tempered gears.

Heat-treated gears appeal to the progressive machine-tool builder. They will make possible the use of gears of smaller section, and while this may not be necessary from the standpoint of weight, as is the case with the motor-car builder, an economy of space is frequently desirable. The greatest advantage is, perhaps, the elimination of repairs and replacements.

The information contained in the foregoing, relating to heat-treatment of gears, is abstracted from a paper read by Mr. J. H. Parker, before the National Machine Tool Builders' Association.

Chapter Thirteen - Testing the Hardness of Metals

Importance of Hardness Tests. — Few properties of iron and steel are of more importance than that of hardness. In some cases, as with a cutting tool or a pressure die, the metal is practically valueless unless it can retain a sharp edge; while in other instances, where the material has to be machined or cut or trued to shape, even a relatively slight increase of hardness is the cause of much inconvenience and expense. In a third class of material a good wearing surface is of prime importance; while, lastly, hardness may often serve as an indication of a degree of brittleness and untrustworthiness which might perhaps be otherwise unsuspected.

Definition of Hardness. — Hardness may be defined as the property of resisting penetration, and, conversely, a hard body is one which, under suitable conditions, readily penetrates a softer material. There are, however, in metals various kinds or manifestations of hardness according to the form of stress to which the metal may be subjected. These include tensile hardness, cutting hardness, abrasion hardness, and elastic hardness; doubtless other varieties could also be recognized when the experimental conditions are modified so as to bring into operation properties of the material in addition to that of simple, or what may be conveniently called mineralogical hardness. This has been defined by Dana as "the resistance offered by a smooth surface to abrasion."

The usual quantitative tests for hardness are static in character, but the conditions are profoundly modified when the penetrating body is moving with greater or less velocity. The resistance to the action of running water, to the effect of a sandblast, or to the pounding of a heavy locomotive on a steel rail, affords examples of what might perhaps for purposes of distinction be called dynamic hardness, which is a branch of the subject that has received little attention.

Simple Methods for Testing Hardness. — Even when men first began to harden steel, they probably sought some method of ascertaining in particular cases whether their object had been accomplished. Perhaps the testing tool was nothing more than a fragment of flint or another piece of steel known to be hard. Certain jewels — as the diamond — are well suited to a process which depends upon scratching. In fact, this process is in common use everywhere even at the present day.

The test by filing is not to be despised as it is easily applied, and if the file is a good one, the results are sufficiently accurate and reliable for a considerable class of work. But the file is an instrument inadequate to the requirements of modern metallurgists and manufacturers. This is true for two reasons: First, the alloy steels seem to possess the property of being able to resist a file, apart from hardness. Thus, a piece of manganese self-hardening tool steel may be, in reality, softer than a specimen of a pure carbon steel, and yet resist the attacks of the file equally well.

In explanation of this phenomenon, it has been suggested that the hard manganese resists the file while the iron substratum remains soft. The combination as a whole would not be so hard, although able to withstand the file. This, however, seems really to involve the proposition that such steel is not a perfect chemical combination, but that particles of manganese are held imbedded in iron or an iron alloy. Perhaps this may be so, but if it is true, then the action of such steel on the file is very similar to that of an emery wheel. The emery itself is very hard, but is held in a matrix that is soft. However, whether we accept this explanation or not, it is doubtful whether we have good reason to contend that a specimen of alloy steel is as hard as a piece of pure carbon steel, merely because it resists the file equally well.

The second objection to the file is that it affords no reliable means of making accurate comparisons between different degrees of hardness. It is sometimes of importance in cases where one element of a machine slides against another to ascertain which of the two is the harder. The difference may be very slight, yet it will readily be granted that this difference might become of importance if lubrication failed, for the harder piece would then cut or wear the softer. If such a contingency is possible, then it is important that the more expensive part shall be the harder. A little reflection will convince one that this principle of associating a harder valuable part with a softer less valuable part has application everywhere in machine construction; but in order to apply this principle widely, it is necessary to be able to determine differences in hardness where these differences are quite small in amount.

Modern Methods for Testing Hardness. — Comparison will be made in the following of four typical methods of measuring hardness. Those selected include the sclerometer, introduced by Prof. Thomas Turner in 1886; the scleroscope, introduced in 1907 by Shore; the form of indentation test adopted by Brinell about 1900; and the drill test introduced by Keep in 1887. Each of these methods has been used in actual works practice, and by various persons other than the inventor, and may thus be regarded as being typical of the particular class of test to which it belongs. Among the many other forms of test, the microsclerometer and wearing tests call for special mention, though to these only incidental reference can be made. The principles underlying the four methods selected for comparison will be briefly described in the following.

Turner's Sclerometer. — In this form of test a weighted diamond point is drawn, once forward and once backward, over the smooth surface of the material to be tested. The hardness number is the weight in grams required to produce a standard scratch. The scratch selected is one which is just visible to the naked eye as a dark line on a bright reflecting surface. It is also the scratch which can just be felt with the edge of a quill when the latter is drawn over the smooth surface at right angles to a series of such scratches produced by regularly increasing weights.

Shore's Scleroscope. — In this instrument, which will subsequently be described in detail, a small cylinder of steel, with a hardened point, is allowed to fall upon the smooth surface of the metal to be tested, and the height of the rebound of the hammer is taken as the measure of hardness. The hammer weighs about 40 grains, the height of the rebound of hardened steel is in the neighborhood of 100 on the scale, or about 6j inches, while the total fall is about 10 inches or 255 millimeters.

Brinell's Test. — In this method, described in detail in succeeding pages, a hardened steel ball is pressed into the smooth surface of the metal so as to make an indentation of a size such as can be conveniently measured under the microscope. The spherical area of the indentation being calculated, and the pressure being known, the stress per unit of area when the ball comes to rest is calculated, and the hardness number obtained. Within certain limits the value obtained is independent of the size of the ball, and of the amount of pressure. In the original tests the steel ball was 10 millimeters (0.394 inch) in diameter, and the pressure was equal to a weight of 3000 kilograms; but a more convenient form of apparatus is now supplied by Mr. Brinell for works tests, while Mr. Stead and Mr. Derihon have introduced small portable instruments.

Keep's Test. — In this form of apparatus a standard steel drill is caused to make a definite number of revolutions while it is pressed with standard force against the specimen to be tested. The hardness is automatically recorded on a diagram on which a dead soft material gives a horizontal line, while a material as hard as the drill itself gives a vertical line, intermediate hardness being represented by the corresponding angle between o and 90 degrees.

Comparison between Testing Methods. — Each form of test has its advantages and its limitations. The sclerometer is cheap, portable, and easily applied, but it is not applicable to materials which do not possess a fairly smooth reflecting surface, and the standard scratch is only definitely recognized after some experience. The Shore test is simple, rapid, and definite for materials for which it is suited, and appears likely to have an important future; but further information is yet needed as to the exact property which is measured by this form of test. As shown by De Fréminville, the result obtained varies somewhat with the size and thickness of the sample, while if the test-piece is supported on a soft material, such as a plasticine, the results are valueless. It should also be pointed out that india-rubber gives a rebound of 23, which is equal to that of mild steel, while light soft pine wood gives a rebound of 40, which is nearly twice as great as that of gray cast iron. Curiously enough, hard wood, like teak, gives a rebound of about 12, while some samples are considerably lower than this.

As illustrating the influence of the support, a sample of exceptionally hard rolled copper, about 0.040 inch in thickness, when supported on a block of hard steel, and tested with the blunt or "magnified" hammer supplied, gave a value of 30, which was increased to 34 when the copper was supported on wood. A sample of brass only gave a value of 17, and yet this brass would scratch the copper, while the copper would not scratch the brass. From these results it would seem that the Shore test is only applicable to a certain class of substances. It appears to test what may be termed the "elastic hardness," and gives high results with metals in the "worked hard" condition. Tests appear to show that good results are, however, obtained with glass and with porcelain, as well, of course, as with most metals.

The Brinell test is especially useful for constructive material; it is easily applied and definite, and is now of all hardness tests the one most employed. It appears to give satisfactory results with wood, but cannot be applied to very brittle materials, such as glass, or to hard minerals. Keep's test is especially suited for castings of all kinds, as it records not merely the surface hardness, but also that of the whole thickness, and gives indications of blowholes, hard streaks, and spongy places. Obviously, it can only be applied to materials the hardness of which is less than that of hardened steel.

Comparison of Results Obtained by Different Testing Methods. — A very important question arises in connection with these various tests — namely, as to whether there is any observed agreement between the results which are arrived at by such entirely different methods. It will be noticed that in each case an arbitrary scale is adopted. If the weights used on the sclerometer had been ounces instead of grams, the hardness numbers would naturally have been different. Similarly, Brinell's tests might have been expressed in tons and inches, or a different weight of hammer and height of scale adopted by Shore. Hence all that can be expected is a proportionality in the results, and if this is ascertained it should be possible to convert values on one scale into results on another.

Metal	Sclerometer	Scleroscope	Brinell
Lead	1.0	1.0	1.0
Tin	2.5	3.0	2.5
Zinc	6.0	7.0	7.5
Copper, soft	8.0	8.0	....
Copper, hard	....	12.0	12.0
Softest iron	15.0	....	14.5
Mild steel	21.0	22.0	16–24
Soft cast iron	21–24	24.0	24.0
Rail steel	24.0	27.0	26–35
Hard cast iron	36.0	40.0	35.0
Hard white iron	72.0	70.0	75.0
Hardened steel	....	95.0	93.0

Table 1. Hardness Scales Compared

An examination of results obtained by the four methods dealt with shows that, for relatively pure metals in their cast or normal condition, there is a general agreement which must be regarded as remarkable. In Table 1 will be found, in the first column, results which were published by Prof. Turner in a paper on the hardness of metals in 1886. In the second column are Prof. Turner's results with the Shore scleroscope, and these figures are in good agreement with those supplied by the maker of the instrument. In the third column are values taken from published results by Mr. Brinell and by Mr. Stead, but the numbers given have been divided by 6, as this figure has been found to suitably reduce the Brinell hardness values for purposes of comparison.

It will be observed that either by accident or design the scale adopted for the scleroscope is, for practical purposes, identical with that of the sclerometer, while Mr. Brinell's values are proportional. The angles in Keep's tests could easily be made to show pretty close agreement with the other values. It would therefore appear that each instrument, with simple and homogeneous substances, must measure one and the same physical property, and give results which are either in actual agreement with, or proportional to, the results obtained by the other forms of tests.

Hardness Tests on Worked or Heat-treated Metals. — In practice, however, the use of relatively pure metals in the unworked or annealed condition is comparatively rare, unless we include in this category wrought iron and mild steel. Carbon steels and special steels consist largely of alloys, the complexity of which is profoundly modified by heat-treatment; while copper, zinc and their alloys are frequently hardened by rolling, drawing or other mechanical treatment.

The very important question therefore arises as to the extent to which the different methods of testing agree in their values for hardened and tempered steel, and for the hardness caused by mechanical treatment. From preliminary observations on the latter point Prof. Turner states as his belief that metal which has been mechanically treated, as with hard-drawn rods or rolled sheets, has its tenacity increased out of proportion to its hardness as measured by a file or cutting tool. The sclerometer shows relatively little difference, for example, between hard-drawn and annealed copper, while the scleroscope shows an exaggerated effect, at all events in some cases. As the Brinell test closely follows the tenacity, it too may be expected to show a marked difference between worked and annealed samples. The result in some cases is likely to be a confusion between elasticity or tenacity on the one hand, and true or mineralogical hardness on the other. For example, a piece of hard-rolled copper may give a greater hardness number than one of mild steel; yet a tool made of mild steel will always cut copper, but no amount of cold-rolling will make copper cut steel. Hence great care is required when hardness values for different materials are compared.

Table II. Percentage of Loss of Hardness of Hardened Steel when tempered to various temperatures, as measured by different hardness testing apparatus

Temperature of Heating, Degrees C.	Brinell Method (0.83 per cent Carbon)	Martens Sclerometer (0.95 per cent Carbon)	Jagger Microsclerometer (0.86 per cent Carbon)	Shore's Method (0.83 per cent Carbon)
160	..	2.5	1.8	3.7
200	13	14.0	5.4	2.7
300	38	41.0	9.1	11.1
400	68	70.6	23.6	33.0*
500	94	87.5	64.0	92.5
600	100	95.7	94.5	100.0

Hardness of Steel in Hardened, Tempered or Annealed Condition. — The question of agreement in reference to the true hardness of a sample of steel in the normal, hardened, tempered, or annealed condition is perhaps of even greater importance. To illustrate the kind of difficulty which arises, reference may be made to some recently published results by E. Maurer, in which samples of steel with varying content of carbon were heated to ascertained temperatures, quenched, and afterward tempered or annealed at given temperatures. The hardness of the samples was then determined. When the tempering heat was 300 degrees C, the loss of hardness in a sample containing 0.83 per cent of carbon was 11.1 per cent by the Shore method, and 38.0 per cent by the Brinell test. A steel with 0.95 per cent carbon tested in a similar manner by Heyn and Bauer with a Martens sclerometer gave a loss of hardness of 41.0 per cent; while lastly Boynton, with a Jagger sclerometer, using a steel with 0.86 per cent carbon, has recorded a loss of hardness on tempering at 300 degrees of only 9.0 per cent.

The question may be put in this way: The steel is suited for making woodworking tools, if properly hardened and tempered; is 300 degrees C. a proper tempering heat? According to the Shore test and the Jagger test the tool should be hard and cut well; but according to the Brinell test and the Martens sclerometer it has lost nearly half its original hardness, and should rapidly lose its cutting edge. Maurer states that everyday experience shows that with this class of tool steel a tempering heat of 300 degrees renders the metal useless for woodworking. The results of the four sets of experiments are given in Table 11.

The values are graphically represented in the engraving, Fig. 1, from which it will be seen that the greatest difference occurs at about 300 degrees C, the loss of quenching hardness due to tempering being about four times as great when tested by the two first methods as compared with the results obtained when the steel is tested by the two latter methods given in the table. Martens and Heyn have recently pointed out that in the Brinell ball test for hardness, the indentations are frequently not circular, and are therefore difficult to measure, and that when testing hard materials the ball itself is appreciably flattened while under load. To diminish these sources of error Martens has introduced a special form of apparatus for measuring the depth of the indentation.

Relation between Hardness and Wear of Steel. — In a paper presented before the International Association for Testing Materials, at its congress in New York, September, 1912, Mr. E. H. Saniter stated, as the results of experiments made, that there is no definite relation between hardness, as measured by the Brinell hardness testing method, and wear. While, in general, a high Brinell hardness number may be expected to indicate a metal which will give better wear, there are so many exceptions that this test for indicating wearing properties would be unreliable. As an example, it was mentioned that Hadfield's manganese steel, which has a low Brinell hardness number, proved the best steel as far as wear is concerned in the wear tests undertaken. However, the relation of either Brinell tests or ordinary wear tests to wear in actual practice is a subject which requires further investigation, and it is not certain that the ordinary type of wear test is a fully reliable indica-

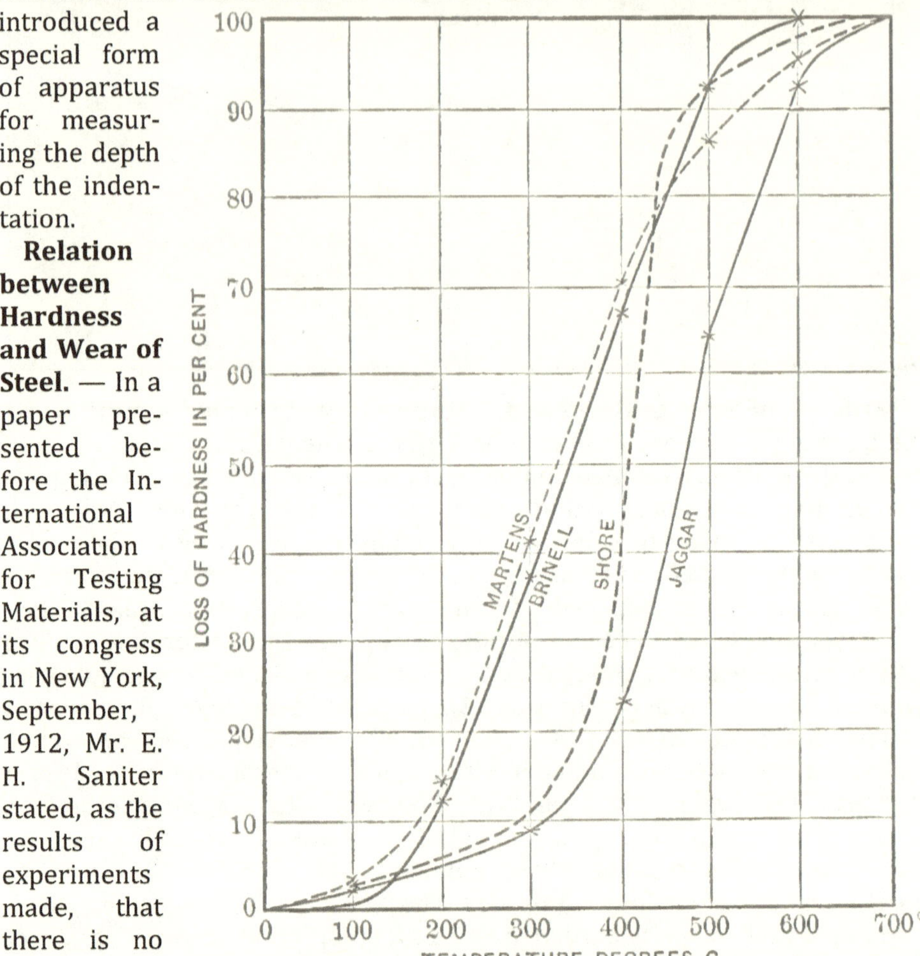

Fig. 1. Diagram showing Percentage of Loss of Hardness in Hardened Steel, when Tempered to Various Temperatures, as measured by Different Hardness Testing Methods

tion of the wear a certain material may give when in use under certain conditions. It would seem, from opinions expressed at the congress for testing materials, that wear tests should be made along different lines according to the actual uses to which the metal is to be put.

The Brinell Method of Testing the Hardness of Metals. — The method of testing the hardness of metals devised by Mr. J. A. Brinell has received very favorable attention from metallurgists in this, as well as in other countries. In 1900 Mr. Brinell, then chief engineer and technical manager of the Fagersta Iron and Steel Works in Sweden, first made public his method of testing the hardness of iron and steel, by submitting it to the Society of Swedish Engineers in Stockholm. At the meeting of the International Congress for Testing Materials in Paris the same year the method attracted general attention, and its merits were duly acknowledged by awarding the inventor with a personal *Grand Prix* at the Paris Exposition.

The method was first described in the English language by Mr. Axel Wahlberg in a paper before the Iron and Steel Institute in 1901. Since then, the practical value of this method has been amply substantiated on various occasions by means of comprehensive tests and investigations undertaken by several distinguished scientists in different countries. In working out his method, Brinell kept in view the necessity of taking into account the requirements that the method must be trustworthy, must be easy to learn and apply, and capable of being used on almost any piece of metal, and particularly, to be used on metal without in any way being destructive to the sample.

Principle of Brinell Method for Testing Hardness of Metals. — The Brinell method, as already mentioned, consists in partly forcing a hardened steel ball into the sample to be tested so as to effect a slight spherical impression, the dimensions of which will then serve as a basis for ascertaining the hardness of the metal. The diameter of the impression is measured, and the spherical area of the concavity calculated. On dividing the amount of pressure required in kilograms for effecting the impression, by the area of the impression in square millimeters, an expression for the hardness of the material tested is obtained, this expression or number being called the *hardness numeral.*

In order to render the results thus obtained by different tests directly comparable with one another, there has been adopted a common standard with regard to the size of ball as well as to the amount of loading. The standard diameter of the ball is 10 millimeters (0.3937 inch) and the pressure 3000 kilograms (6614 pounds) in the case of iron and steel, while in the case of softer metals a pressure of 500 kilograms (1102 pounds) is used. Any variation either in the size of the ball or the amount of loading will be apt to occasion more or less confusion without there being any advantage to compensate for such inconvenience. Besides, making any comparisons between results thus obtained in a different manner would be more or less troublesome, and compHcated calculations would be required.

The diameter of the impression is measured by means of a microscope of suitable construction, and the hardness numeral may be obtained without calculation directly from the table given herewith, worked out for the standard diameter of ball and pressures mentioned. The formulas employed in the calculation of this table are as follows:

$$y = 2\pi r\left(r - \sqrt{r^2 - R^2}\right),$$
$$H = \frac{K}{y}$$

in which formulas

r — radius of ball in millimeters;

R = radius of depression in millimeters;

y = superficial area of depression in square millimeters;

K = pressure on ball in kilograms;

H = hardness numeral.

Suppose, for instance, that the radius of the ball equals 5 millimeters (0.1968 inch), and that the test is undertaken on a piece of steel, the pressure consequently applied being 3000 kilograms (6614 pounds). Assuming that we found the radius of the depression equal to 2 millimeters (0.07874 inch) by measurement, we have:

$$2\pi \times 5\left(5 - \sqrt{25 - 4}\right) = 13.13 = y,$$

and

$$\frac{3000}{13.13} = 228 = H,$$

which as we see agrees with the figure given in our table for 4 millimeters diameter of impression.

Relation between Hardness of Materials and Ultimate Strength. — It has been pointed out by Mr. Brinell himself that this method of testing the hardness of metals offers a most ready and convenient means of ascertaining within close limits the ultimate strength of iron and steel. This, in fact, is one of the most interesting and important results of this method of measuring hardness. In order to determine the ultimate strength of iron and steel, it is only necessary to establish a constant coefficient determined by experiments which serves as a factor by which the hardness numerals are multiplied, the product being the ultimate strength. Rather comprehensive experiments were undertaken with a considerable number of specimens of annealed material obtained from various steel works, for the purpose of establishing the coefficients, by the present director of the Office for Testing Materials of the Royal Technical Institution at Stockholm. The results obtained were as follows: For hardness numerals below 175, when the impression is effected transversely to the rolling direction, the coefficient equals 0.362; when the impression is effected in the rolling direction, the coefficient equals 0.354.

For hardness numerals above 175, when the impression is effected transversely to the rolling direction, the coefficient equals 0.344; when the impression is effected in the rolling direction, the coefficient equals 0.324.

Diameter of Impression, mm.	Hardness Numeral		Diameter of Impression, mm.	Hardness Numeral		Diameter of Impression, mm.	Hardness Numeral		Diameter of Impression, mm.	Hardness Numeral		Diameter of Impression, mm.	Hardness Numeral	
	Pressure in Kilograms			Pressure in Kilograms			Pressure in Kilograms			Pressure in Kilograms			Pressure in Kilograms	
	3000	500		3000	500		3000	500		3000	500		3000	500
2.00	946	158	3.00	418	70	4.00	228	38.0	5.00	143	23.8	6.00	95	15.9
2.05	898	150	3.05	402	67	4.05	223	37.0	5.05	140	23.3	6.05	94	15.6
2.10	857	143	3.10	387	65	4.10	217	36.0	5.10	137	22.8	6.10	92	15.3
2.15	817	136	3.15	375	63	4.15	212	35.0	5.15	134	22.3	6.15	90	15.1
2.20	782	130	3.20	364	61	4.20	207	34.5	5.20	131	21.8	6.20	89	14.8
2.25	744	124	3.25	351	59	4.25	202	33.6	5.25	128	21.5	6.25	87	14.5
2.30	713	119	3.30	340	57	4.30	196	32.6	5.30	126	21.0	6.30	86	14.3
2.35	683	114	3.35	332	55	4.35	192	32.0	5.35	124	20.6	6.35	84	14.0
2.40	652	109	3.40	321	54	4.40	187	31.2	5.40	121	20.1	6.40	82	13.8
2.45	627	105	3.45	311	52	4.45	183	30.4	5.45	118	19.7	6.45	81	13.5
2.50	600	100	3.50	302	50	4.50	179	29.7	5.50	116	19.3	6.50	80	13.3
2.55	578	96	3.55	293	49	4.55	174	29.1	5.55	114	19.0	6.55	79	13.1
2.60	555	93	3.60	286	48	4.60	170	28.4	5.60	112	18.6	6.60	77	12.8
2.65	532	89	3.65	277	46	4.65	166	27.8	5.65	109	18.2	6.65	76	12.6
2.70	512	86	3.70	269	45	4.70	163	27.2	5.70	107	17.8	6.70	74	12.4
2.75	495	83	3.75	262	44	4.75	159	26.5	5.75	105	17.5	6.75	73	12.2
2.80	477	80	3.80	255	43	4.80	156	25.9	5.80	103	17.2	6.80	71	11.9
2.85	460	77	3.85	248	41	4.85	153	25.4	5.85	101	16.9	6.85	70	11.7
2.90	444	74	3.90	241	40	4.90	149	24.9	5.90	99	16.6	6.90	69	11.5
2.95	430	73	3.95	235	39	4.95	146	24.4	5.95	97	16.2	6.95	68	11.3

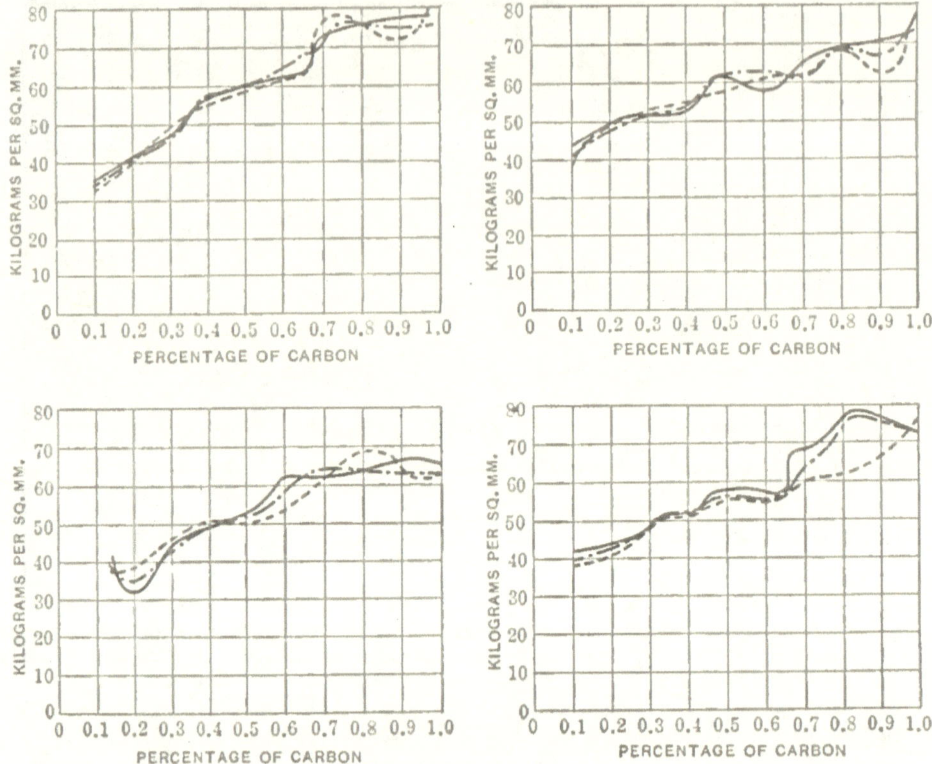

Fig. 2. Diagram showing Relation between Results obtained by Various Methods of Ascertaining the Ultimate Strength of Materials

If the hardness numerals are multiplied by these coefficients, the result obtained will be the ultimate tensile strength of the material in kilograms per square millimeter. It is evident that coefficients can easily be worked out so that if the hardness numerals be multiplied by these the strength could be obtained in pounds per square inch.

Suppose, for instance, that a test of annealed steel by means of the Brinell ball test gave an impression of a diameter of 4.6 millimeters. Then the hardness numeral, according to our table, would be 170, and the ultimate tensile strength consequently 0.362 x 170 = 61.5 kilograms per square millimeter, provided the impression was effected transversely to the rolling direction.

In Fig. 2 are shown a number of diagrams which indicate the results obtained from the tests undertaken to ascertain the coefficients given. In these diagrams the full heavy line indicates the tensile strength of the material, as calculated from the ball tests in the rolling direction. The dotted lines indicate the strength as calculated from the ball tests in a transverse direction, and the "dash-dotted" lines show the actual tensile strength of the material as ascertained by ordinary methods for determining this value. It is interesting to note how closely the three curves agree with one another, and considering the general uncertainty and variation met with when testing the same kind of material for tensile strength by the ordinary methods, it is safe to say that the ball test method comes nearly as close to the actual results as does any other method used. Especially within the range of the lower rates of carbon, or up to 0.5 per cent, or in other words, within the range of all ordinary construction materials, the coincidents are, in fact, so very nearly perfect as to be amply sufficient to satisfy all practical requirements.

In the case of any steel, whether it be annealed or not, that has been submitted to some further treatment of any other kind than annealing, such as cold working, etc., or in the case of any special steel, there would be other coefficients needed which would then also be ascertained by experiments. The same coefficient, however, will hold true for the same kind of material having been subjected to the same treatment. Thus, the ball testing method for strength is equally satisfactory, and far more convenient, in all cases where the rupture test would be applied. One of the greatest advantages of the Brinell method is that in the case of a large number of objects being required to be tested, each one of the objects can be tested without demolition, and without the trouble of preparing test bars.

Practical Constants for Relation between Hardness and Strength. — In a paper entitled "Researches on the Hardness of Steel," read by Capt. C. Grard before the Congress of the International Association for Testing Materials, held in New York City, September, 1912, the relation between hardness and tensile strength, as determined in recent investigations, was exhaustively dealt with, and the following coefficients were given for different grades of steel:

Steels, extra soft	K = 0.360
Steels, soft and semi-hard	K = 0.355
Steels, semi-hard	K = 0.353
Steels, hard	K = 0.349

It will be seen that these coefficients differ by but a slight amount and it was suggested by another member of the congress of testing materials that a uniform constant be adopted by the International Association for Testing Materials, which would be used for calculating the tensile strength directly from the results of the hardness tests.

Accuracy of Brinell Test. — In a paper read before the Congress of the International Association for Testing Materials held in New York City, September, 1912, an investigation into the accuracy possible with the Brinell hardness testing method was recorded. From this investigation it appears that when commercial apparatus, as ordinarily used for making the Brinell test, is employed, and the test is carried out with ordinary care and precaution, it is reliable within an error of five Brinell units above or below the actual hardness. In other words, if the hardness of two pieces of metal is tested, and the difference on the Brinell scale is more than ten hardness units, it is certain that there is an absolute difference in the hardness of the pieces tested. With regard to the conditions under which the tests should be made, it may be stated that the pressure should be gradually applied during a time of two minutes or more, and the pressure should be kept on the test piece for a period of at least from two to five minutes.

The Time Element in Hardness Tests by the Brinell System. — A diagram indicating the effect of the time element in hardness tests made by the Brinell method is shown in Fig. 3. The tests upon which this diagram is based were made by the German Glyco Metal Co. On the lower scale is given the time, in minutes, during which the pressure on the metal was permitted to act, while the scale on the left-hand side gives the hardness numerals according to the Brinell hardness scale. It will be noted that the longer the pressure was permitted to act, the greater was the impression made on the metal, so that a lower hardness numeral resulted. Two sets of curves are shown, one with the metal heated to 176 degrees F., and one with the metal at 68 degrees F. It is interesting to note that the curves in each set are almost parallel, except in one case, thus indicating that for *comparative* purposes the Brinell test is accurate no matter what the length of duration of pressure, provided, of course, that the various samples tested are all subjected to pressure for the same length of time.

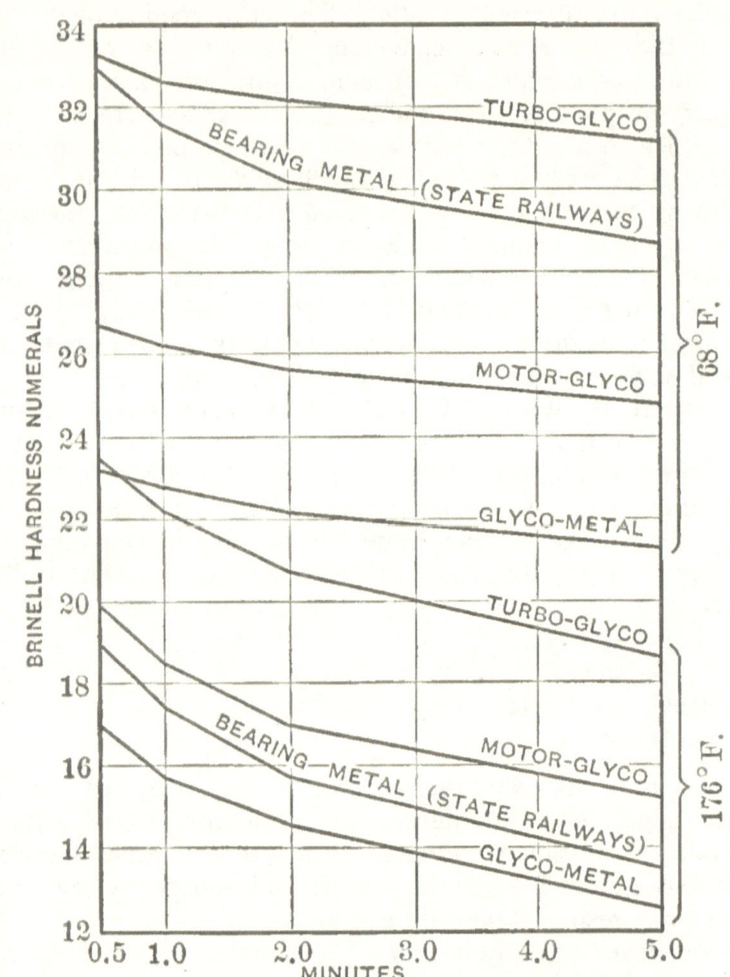

Fig. 3. Curves showing the Variation in Results obtained by Hardness Tests of Varying Duration

It is also interesting to note the difference in the hardness of the metal brought about by the change in temperature. It will be seen that at the higher temperature its hardness numeral is not only less than at the lower temperature, as, of course, would be expected, but when the pressure on the metal is permitted to remain for a longer time, the metal apparently gives way much easier to continuous pressure when heated than when at a lower temperature. Tests of this kind should be of great value in determining the relative value of bearing metals which for long periods are to be subjected to heavy pressures under increasing temperatures. A new factor in hardness testing is also introduced, which the Brinell method is particularly adapted to measure, *viz.*, the power of resistance of various metals to *continuous* pressure, a factor which may be found to vary considerably for different materials.

Application of the Brinell Ball Test Method. — Summarizing what has been said in the previous discussion, and adding some other important points, we may state the various uses for which the Brinell ball test method may be applied, outside of the direct test of the hardness of construction materials and the calculation from this test of the ultimate strength of the materials, as follows:

1. Determining the carbon content in iron and steel.
2. Examining various manufactured goods and objects, such as rails, tires, projectiles, armor plates, guns, gun barrels, structural materials, etc., without damage to the object tested.
3. Ascertaining the quality of the material in finished pieces and fragments of machinery, even in such cases when no specimen bars are obtainable for undertaking ordinary tensile tests.
4. Ascertaining the effects of annealing and hardening of steel.
5. Ascertaining the homogeneity of hardening in any manufactured articles of hardened steel.
6. Ascertaining the hardening power of various quenching liquids and the influence of temperature of such liquids on the hardening results.
7. Ascertaining the effect of cold working on various materials.

Machines for Testing the Hardness of Metals by the Brinell Method. — The method of applying the Brinell ball test was at first only possible in those establishments where a tensile testing machine was installed. As these machines are rather expensive, the use of the ball test method was limited. For this reason a Swedish firm, Aktiebolaget Alpha, Stockholm, Sweden, has designed and placed on the market a compact machine specially intended for making hardness tests. This machine, the most important working mechanism of which is shown in Fig. 4, consists of a hydraulic press acting downward, the lower part of the piston being fitted with a 10-millimeter steel ball *k* by means of which the impression is to be effected in the surface of the specimen or object to be tested. This object is placed on a support (not shown) which is vertically adjustable by means of a hand-wheel, while at the same time it can be inclined sideways when this is needed on account of the irregular shape of the part tested. The whole apparatus is solidly mounted on a cast-iron stand. The pressure is effected by means of a small hand pump, and the amount of pressure can be read off directly in kilograms on the pressure gage mounted at the top of the machine.

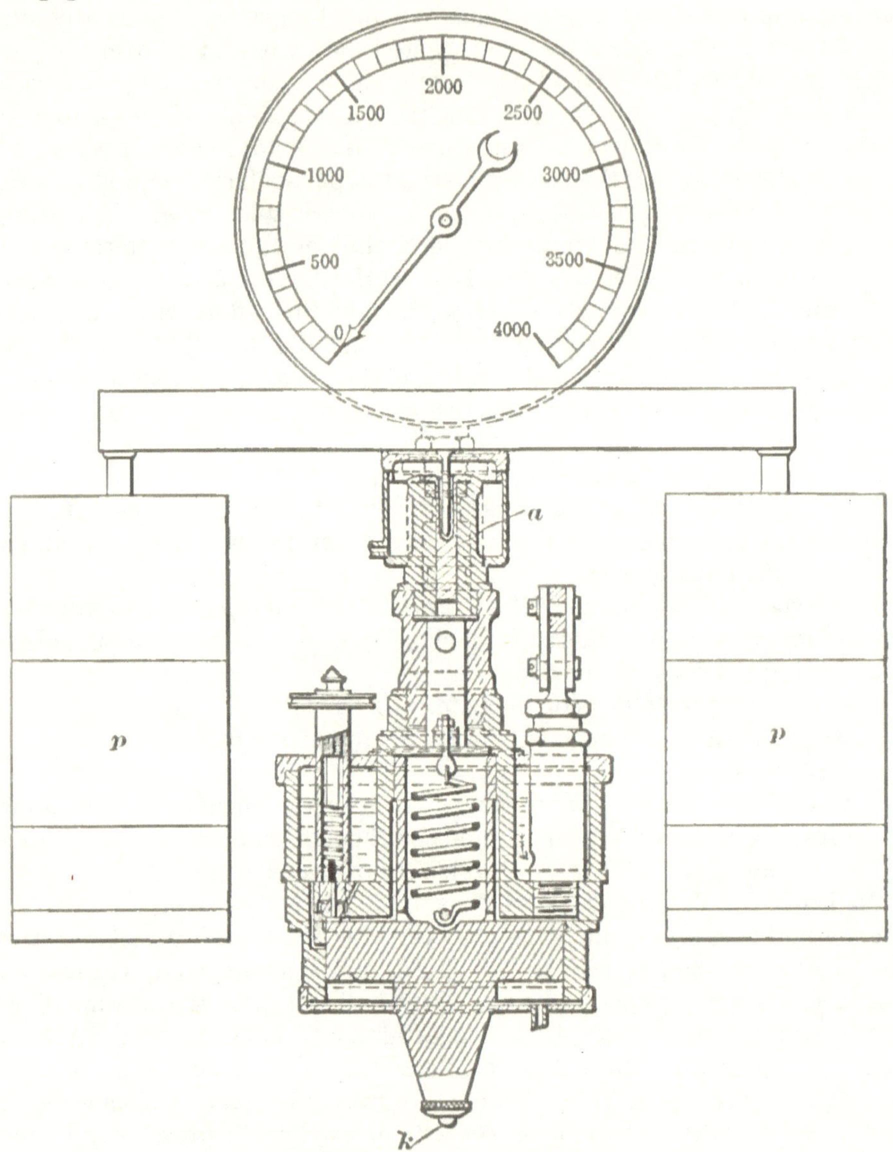

Fig. 4. Section of Working Mechanism of a Hardness Testing Machine employing the Brinell Principle

In order to insure against any eventual non-working of the manometer, this machine is fitted with a special contrivance purporting to control in a most infallible manner the indications of that apparatus, while at the same time serving to prevent any excess of pressure beyond the exact amount needed according to the case. This controlling apparatus consists of a smaller cylinder, *a*, directly communicating with the press-cylinder. On being loaded with weights corresponding to the amount of pressure required, the piston in this cylinder will be pushed upward by the pressure effected within the press-cylinder at the very moment when the requisite testing pressure is attained. Owing to this additional device, there can thus be no question whatever of any mistake or any errors as to the testing results, that might eventually be due to the manometer getting out of order.

Method of Performing the Ball Test. — The test specimen must be perfectly plane on the very spot where the impression is to be made. It is then placed on the support, which, as mentioned, is adjusted by means of a hand-wheel so as to come into contact with the ball k. A few slow strokes of the hand pump will then cause the pressure needed to force the ball downward, and a slight impression will be obtained in the object tested, but as soon as the requisite amount of pressure has been attained, the upper piston is pushed with the controlling apparatus upward, as previously described. On testing specimens of iron and steel, the pressure is maintained on the specimen for 2 minutes, but in the case of softer materials for at least 5 minutes. After the elapse of this time, the pressure is released, and the contact between the ball and the sample will cease. A spiral spring fitted within the cylinder, and being just of sufficient strength to overcome the weight of the press piston, pulls the same upward into its former position, while forcing the liquid back into its cistern. The diameter of the impression effected by the ball is then measured by a microscope, which is specially constructed for this purpose, the results obtained by this measurement being exact within 0.05 millimeter (0.002 inch).

Another type of machine is designed for special tests in which very high pressures are required. The ball in this machine is 19 millimeters (0.748 inch) in diameter, and the pressures employed vary from 3 to 50 tons. The construction and operation of the larger machine are otherwise exactly the same as that of the machine in Fig. 4.

Derihon Portable Form of Brinell Hardness Testing Machine. — The only disadvantage of the Brinell method for general practical use lies in the apparatus required and the comparative slowness of the operation. The apparatus just described, constructed in the form of a small hydraulic press, operated by hand, is rather heavy. It is evident that with such an apparatus, the work has to be brought to the machine, so that its regular use for inspection purposes in different parts of a manufacturing plant is impracticable.

The foregoing considerations lend interest to the apparatus shown in Fig. 5, which was devised by the Derihon Steel Works of Loucin-lez-Liége, Belgium. This firm is an important manufacturer of high-grade drop forgings for automobile work, and it originally developed the machine for use in its own plant. As may be seen, the pressure is applied by a hand-operated screw, and the press is small enough to be perfectly portable, weighing only about 12 pounds.

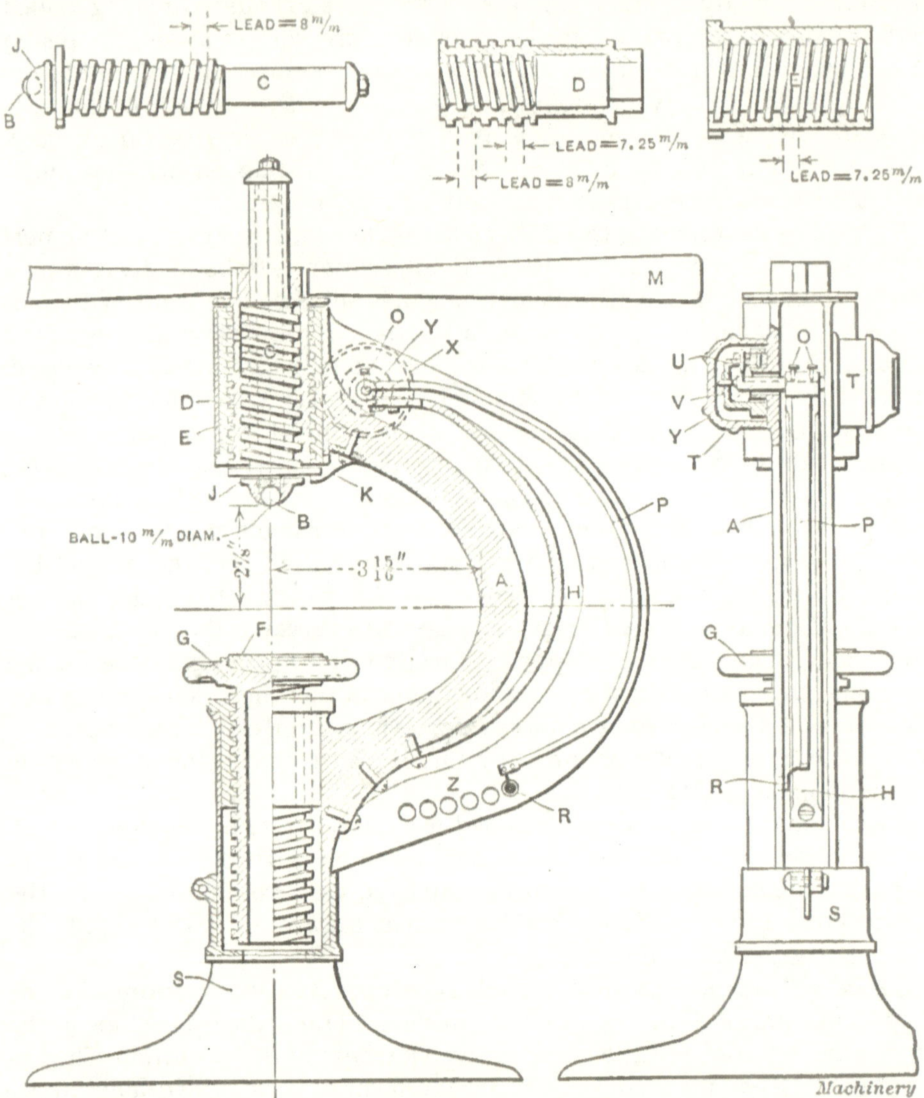

Fig. 5. Section of the Derihon Hardness Testing Apparatus employing the Brinell Principle

The sectional view in Fig. 5 shows clearly the action of the apparatus. The work is placed on the platen *F*, which rests in a spherical seat on top of adjusting screw *G*. By means of this self-adjusting seat, the work gets an even bearing and a direct pressure, even though its under surface may be quite out of true. The purpose of the adjusting screw *G* is, of course, to give a rapid adjustment for the thickness of the work. It will take in about 3 inches as shown. The thread of *G*, while of coarse pitch, still lies within the angle of repose, so that it is not disturbed when the pressure is applied by levers *M*.

A differential screw mechanism is used for applying the pressure. This mechanism consists of the double handle *M*, keyed to the sleeve *D*, which is threaded into the stationary nut *E*, and onto the ram *C*; this latter is kept from revolving by a stop *K*, which enters a slot cut in the flange, and is provided with a threaded chuck for holding the ball *B*. Sleeve *D* is, it will be seen, the only revolving member of this differential screw. The thread on the inside has a lead of 8 millimeters, while that on the outside has a lead of 7.25 millimeters. This gives an advance of 8 - 7.25 = 0.75 millimeters, or about 0.03 inch per revolution of the levers *M*. The advantage of this construction is, of course, that it gives the effect of a fine thread with a lead 0.03 inch, with the use of threads coarse enough to withstand the great strain to which they are subjected in tightening down the work.

The most ingenious feature of the mechanism is the provision made for gaging the pressure with which the ball is forced into the work. By the arrangement used, the frame *A* of the press itself serves as the spring by which the pressure is measured. As the ball is forced into the work with greater and greater pressure, the frame *A* is deflected and *B* and *F* spring apart. Arm *H* is screwed to *A* at its lower end as shown, but is free at its upper end. Here it is provided with a bearing point *X* close to fulcrum *Y* of pointer *P*. As the frame *A* springs under the pressure, *H*, being free at the upper end, remains undistorted and stationary, while pivot *Y* rises. As pivot *Y* rises, lever *P* swings downward, since it rests only on the point *X* of stationary arm *H*. The lower end of lever *P* is provided with a thin metal disk *R*, which is thus swung in an arc of a circle about center *Y*. A series of holes *Z*, bored in the side of the frame, permit the position of this disk to be seen. These holes are so calibrated that each reads to a definite number of kilograms of pressure when disk *R* is centered with it. Under the extreme tension, when central with the left-hand hole, the reading shows the application of a pressure of 3000 kilograms or 6614 pounds.

The construction of the pivot joint at *Y* is interesting, and is best shown at the right of the engraving in Fig. 5. The hub of pointer *P* is clamped to *Y* by two set-screws *O*, which set in a V-slot cut in *Y*. Two caps *T* are screwed on at either side to protect the bearings of pivot *Y*. Set-screws *V* are adjusted to take up the end motion of *F*; they do not, however, restrain it in any direction other than the longitudinal. The real bearing is furnished by the points of screws *U* (one at each end) in the bottoms of the V-grooves. These furnish a knife edge or, rather, point support, which gives the utmost freedom and sensitiveness of movement to the pointer *P* and the indicating disk *R*. The various adjustments in connection with this bearing, and the various contact points in the lever system, will be clearly understood from the engraving.

The simplicity of the operation of this device will be immediately appreciated. The object to be tested is placed on the platen, the adjusting screw is run up until the work makes contact with the ball, and then the handles are revolved until the indicating disk is centered with the particular hole in the frame which shows that the standard pressure has been reached. Handles *M* are then screwed back again, the work is removed, and the diameter of the impression in millimeters measured. This gives the hardness number directly. The whole operation is evidently one of seconds only.

The Ballentine Hardness Testing Device. — In the Brinell hardness testing method the measuring of the dimensions of the indentation is more or less difficult to accomplish accurately, and the method requires special instruments for obtaining the indentation, and for measuring the amount of depression in the metal tested. In order to overcome this difficulty, a means known as the Ballentine method and apparatus for quickly and accurately determining the resistance to indentation of a material has been devised and constructed.

The method employed in the Ballentine device consists in allowing a hammer of specified weight to fall through a specified height on an anvil to which is connected a pin which rests on the specimen to be tested. An indentation in the material is obtained, but the resistance encountered, instead of the dimensions of the indentation, is measured. This resistance is measured by the blow of the hammer being transmitted to the test pin through a soft metal recording disk located at the lower end of the hammer. This disk affords a constant resistance to deformation, and will be indented to a depth varying in proportion to the resistance the pin encounters in indenting the material tested. The recording disk is usually made from lead.

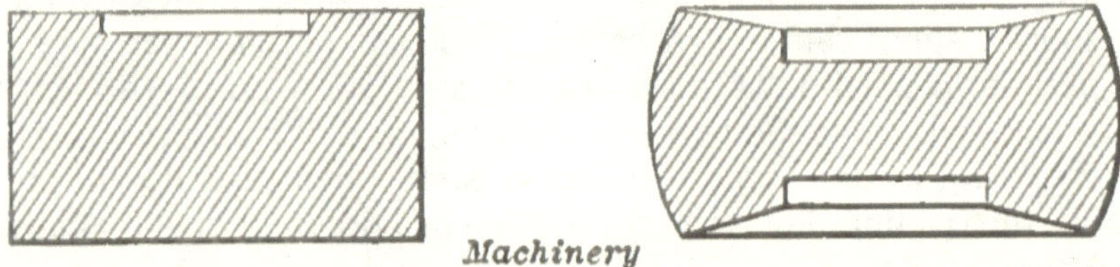

Fig. 6. Lead Recording Disk used in the Ballentine Device, before and after Test

Fig. 7 shows a sectional view of the apparatus, which consists of a guide tube encasing the drop hammer which at the lower end is provided with a small anvil to which is clamped a lead disk. At the upper end the hammer is held at the top of the tube by a spring latch. At the lower end of the tube a test pin holder is located, in which are inserted the test pins for testing the various materials. The upper end of the test pin holder is provided with an anvil of the same diameter as the one on the lower end of the hammer. A small spirit level is inserted in the top of the tube for leveling the apparatus, and two small slots are cut in the guide tube for inserting and removing the recording disks. The apparatus can be used to test all materials which can be ordinarily machined by steel cutting tools, but cannot be used for hardened steel and similar materials which are too hard to be indented in this manner. Two test pins are provided, one for soft materials such as lead and babbitt metals, and another for harder materials such as iron and steel. The pin for hard materials is very short and small in diameter, while the pin for soft materials is longer and larger in diameter.

The testing can be made either on small test specimens or directly on large parts in process of manufacture, the great advantage of this hardness tester being that it is entirely self-contained and well adapted for either laboratory or general shop use. To make a test it is only necessary to smooth off a surface on the specimen to be tested, and clamp it firmly to some rigid body.

In Fig. 6 is shown a lead recording disk before and after the test. These disks are made within 0.0015 inch of nominal size from a material as nearly of uniform density and hardness as obtainable. The disk is measured with a micrometer before being placed on the drop hammer. When the test has been made, the thickness of the metal between the two recording anvils is again measured, and the difference between the two dimensions will indicate the resistance to indentation or the hardness of the material tested. If, for instance, the disk measured 0.300 inch before the test, and 0.156 inch after test, the difference, 0.144 inch, indicates the hardness of the material, and this hardness would be known as No. 144.

Principle of the Shore Scleroscope. — The Shore scleroscope is an instrument in a measure dependent on sensitive touch; or, in other words, it feels the substance much the same as the human fingers. When we touch two or more objects, as, for instance, an orange and an apple, we know that the orange is softer because it yields under pressure more than the apple. We are powerless to measure the hardness of any object that is harder than the fingertips, and there is no way of telling how hard it may be by finger pressure alone.

The sensitive touch of the scleroscope is produced by a tiny hammer dropping from a height of about ten inches onto the metal, hardened steel, etc., which it penetrates slightly. The hammer moves freely, yet snugly, within a glass tube, and weighs about 40 grains. Its striking point consists of an inserted diamond of rare cleavage formation, annealed sufficiently to withstand shocks. This jeweled point is slightly convex and has an area of from about 0.010 to 0.025 square inch. When the plunger strikes the metal to be tested, it reacts or rebounds. The height of this rebound is read on a graduated scale, and an accurate determination of the quantitative hardness of the piece under test is thus obtained.

In the first experiments a steel ball was used as the hammer, but the results were only partially satisfactory. In fact, the inventor was well-nigh on the point of giving up the method when he met the French expert on metals. Dr. Herault. Following out certain of his suggestions, the inventor succeeded in producing a satisfactory instrument for the testing of hardness. The difficulty with the ball-shaped hammer was that it was incapable of striking a

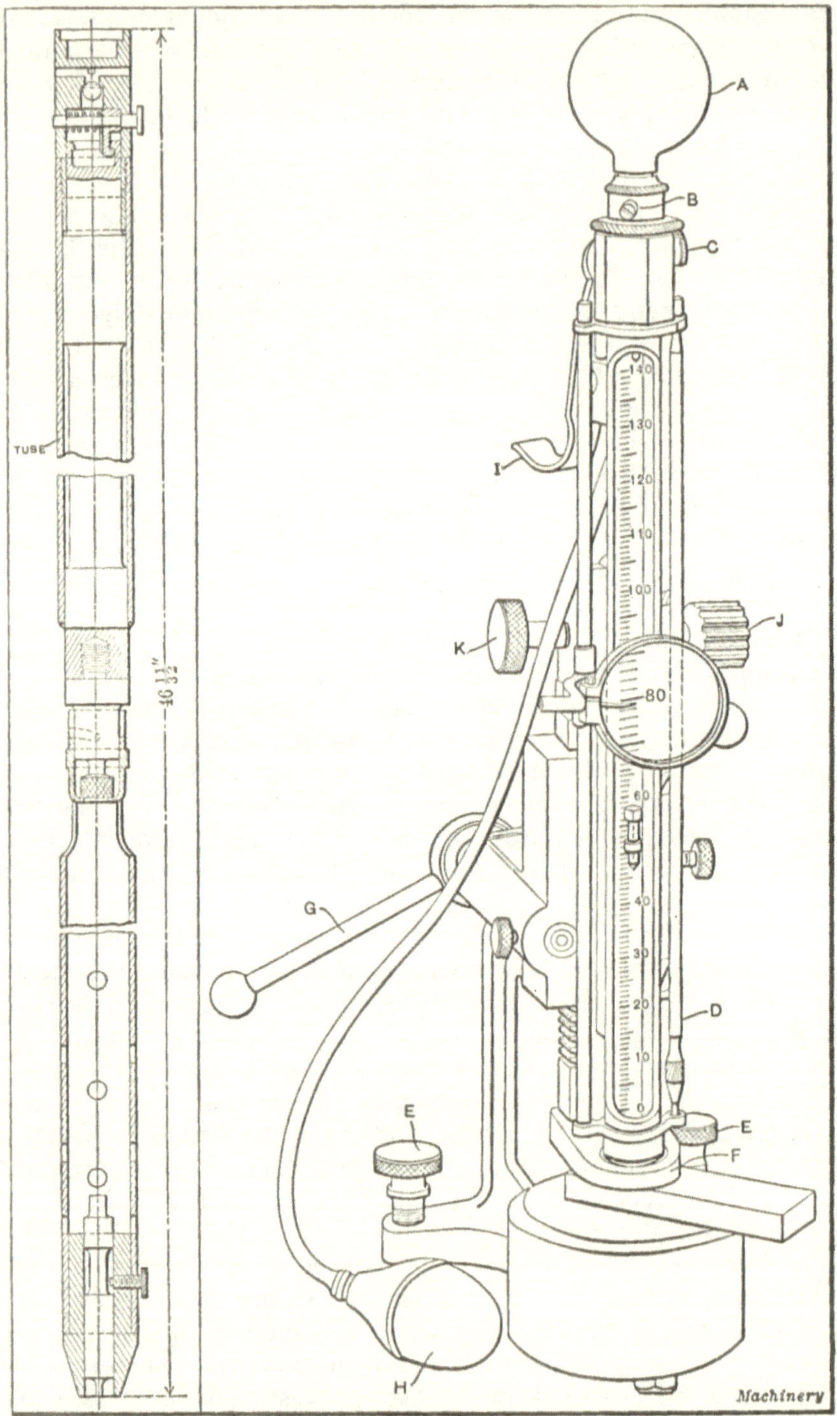

Fig. 7. Ballentine Testing Device

Fig. 8. View showing the Construction of the Shore Hardness Testing Device known as the Scleroscope

sufficiently hard blow to get adequate results, especially with hardened tool steel, so the area of contact was reduced, although the weight was kept large in comparison.

Hardness vs. Elasticity. — When the hammer of the scleroscope is allowed to drop with no other force than its own weight, and the point is so flat that absolutely no impression is made on the surface of very hard steel, then the rebound will be about 90 per cent of the fall. This phenomenon is known as the elasticity of solid bodies. Now, since hardness is resistance to penetration, in its clearest definition, it stands to reason that the point of the hammer must be somewhat reduced and rounded. Therefore the relation between the weight of the hammer and its point should be such that when it drops on hardened steel, a permanent impression must always be made, so that if we had not the rebound to go by, the microscope would still show the values. When the area of the hammer is thus reduced enough to make a permanent impression, a certain amount of the energy stored in the hammer is utilized in doing work. This overcomes the tendency of the metal to resist penetration, depending on how hard it is, or the resistance it offers, and naturally it must rebound considerably less. The hammer always delivers a blow of exactly the same force. If now we get a rebound of 75 per cent on very hard steel, we know that 15 per cent of the hammer's energy was spent in its efforts to overcome the resistance of the steel before it had a chance to react and repel the missile.

Description of the Shore Instrument. — While the absolute weight of the entire hammer is little, it is very great relative to the striking area. The hammer has a cylindrical body and is guided in its fall by a glass tube. Great difficulty has been experienced in obtaining tubes with a sufficiently perfect bore. There seems to be no commercial method of manufacturing such tubes, and the method of "test-and-reject" is therefore employed, resulting in a very great amount of waste.

The glass tube is secured to a frame in a vertical position with the lower end open. The operation of the instrument is very simple. When the hammer is to be raised to the top, the bulb *A*, Fig. 8, is pressed and then suddenly released. This sucks up the jeweled plunger hammer referred to so that it may be caught by a hook which is suspended exactly central in the glass tube and engages with an internal groove on the top of the hammer. Adjusting screws *B* for the hook and its spring are contained in the removable knurled cap. *C* is a cylinder and piston for releasing the hook and hammer by bulb *H* whenever a test is to be made; *I* is a hook which is pressed at the same time and which opens a valve letting in the air and thus preventing the occurrence of a vacuum when the hammer drops. At *J* is shown a pinion knob for moving the instrument up and down independently of the heavy rack and clamp *F* actuated by the lever *G*. At *E* are shown leveling screws and at *D* a plumbrod.

Application of the Shore Instrument. — When small pieces are to be tested, the scleroscope, as shown in Fig. 8, self-contained with its clamp and anvil, is employed. In using the instrument with the stand the specimen is placed on the table or secured in a holder. It is necessary that the actual point tested should be clean and horizontal and that the piece should be firmly held. If necessary to test more than once, the piece should be slightly moved so as to expose a fresh point to the hammer. The indentation made is, however, very minute, so that several are usually unobjectionable.

When the ends of rods, drills, and many other tools are to be tested, they are clamped in a bench vise, and a swinging arm is employed; the instrument is removed from its post on the clamp frame by knurled set-screw *K*, and is attached in the same way to the post on the swinging arm. A kind of female dovetailed finger ring attached to the clamp on the dove-tail rack bar of the instrument is provided for use in free-hand testing on very large floor work, on parts of machinery being assembled, or on the stock rack, etc. From what has been said, it is apparent that the apparatus is of universal application.

The Shore Hardness Scale. — Instead of dividing the whole length of the fall of the hammer into a scale consisting of 100 divisions, the figure 100 is carried down to a point representing about 68 per cent of the total height of the scale as shown in Fig. 8. This was not an arbitrary provision, but was adopted after consultation with leading metallurgists, one of whom was Dr. Paul Herault, of aluminum and electric steel making fame, of France. These authorities agreed that in the scleroscope hardened steel of average hardness should be taken as the standard with which all other less hard metals should be compared; 100 is the average hardness of hardened steel; 90 is a low value, while 110 is a high value.

The scale, therefore, makes it an easy matter to compare the various metals, no matter what their hardness is, and the rebound of the hammer is, therefore, measured against a scale graduated from 0 to 140. This scale is secured in position back of the glass tube. To aid in reading the rebound, a magnifying glass is supplied. After some practice the assistance of this glass may be dispensed with. However, when used, it is secured in such a position as to cover the probable region of the expected reading. The rod to the left of the tube is the support to which the magnifying attachment is secured and along which it is adjusted. The rod to the right of the tube is a plum rod; it swings freely from a point of attachment above, and enables the operator to keep the glass tube vertical.

With the scale graduated from 0 to 140, with hardened steel at or near 100, the hardness of all ordinary materials can be measured. Porcelain and glass, of course, have a higher hardness number than hardened steel, while unhardened steels, brass, zinc and lead have gradually lower degrees of hardness. Unhammered or unrolled lead produces a rebound of only two graduations.

Some Uses for the Scleroscope. — One of the results of the introduction of scientific methods of precise quantitative measurement of hardness promises to be in the determination of the relation of the cutting tool to the work to be machined. We are all aware that the tool must be harder; but how much harder? And how express this relation in intelligible language? The scleroscope, it is hoped, will afford a fairly definite answer to this problem. The law has been laid down that the comparative hardness between tool and work, as determined by scleroscope readings, should be in the ratio of 3 to 1 or 4 to 1, in order to secure the best commercial results.

For example, take the case of work to be machined consisting of a 1 per cent carbon tool steel. Unannealed, such steel is found upon testing to have a hardness varying from 40 to 45 points. According to the above law, the cutting tool should be at least about 120 to 135 points hard; but the same steel, properly annealed, is only about 31 points hard. Consequently, it is not difficult to find a suitable material for the cutting tools. A good quality of carbon tool steel, well hardened, has a hardness of from 95 to 110, and is consequently suitable to cut material of a hardness of 31. Now if this principle as to relative hardness can be thoroughly established for all kinds of metals, an element of certainty will be introduced into shop practice.

Again, it is, of course, to be expected that if two metal parts wear or rub against each other, the harder of the two will cut the softer, whether the difference is small or great, so that it is often important to know whether the more expensive part is really the harder. The scleroscope would seem to afford a means of determining with precision slight differences in hardness, thus enabling the manufacturer to assemble contacting moving parts on the principle of a harder expensive piece in association with a softer cheaper one. Thus in an electrical repair shop, instances may readily be found of the steel shaft cut by the brass box, the box cut by the shaft, and a fairly even wear of both. From an economical point of view, it is much better to have the brasses worn than the shaft, and with such an instrument as the scleroscope it would be possible to predetermine this economically better result. It would seem an easy matter for an automobile manufacturer, say, so to specify the hardness of the gears used, that the gear manufacturer could supply him with a uniform product.

An important application of quantitative hardness tests would appear to be in connection with high-speed steels. Such steels disclose, upon testing with this instrument, a hardness varying from 80 to 105. This is at ordinary temperatures, however, and shows scarcely as high a degree of hardness as the best of the pure carbon steels. The effectiveness of high-speed steels depends largely upon the fact that at temperatures of from 600 to 1000 degrees F., at which carbon steels would lose their temper, they retain a high degree of hardness, amounting, say, to 75 on the scleroscope scale. This is sufficient — following the principle of 3 to 1 — to do heavy machining on annealed machine steel having a hardness of 25 on the same scale. But if the heat developed by high speed and heavy cuts succeeds in lowering the hardness of the high-speed steel of the tool much lower than 75, then it is no longer an effective tool. It becomes of importance then to test high-speed steels for their effectiveness under temperature conditions obtaining in actual service. It is a comparatively unimportant matter to know that a certain tool of high-speed steel is very hard when cold; what is its condition when hot? By heating the tool to the required temperature, and then testing with the scleroscope, this condition may be determined. Thus the real effectiveness of the high-speed steels may be determined in advance of their use, or even of their purchase.

Amount of Pressure for Indentation. — When the hammer falls through a height of ten inches onto hardened steel, it will deliver a striking energy equal to about 20,000 times its own weight, acting through a very short space, of course. With a hammer weighing about 40 grains, and an indentation of, say, 0.002 inch depth, a working pressure of about 100 pounds is obtained. This force, acting on a convex point of 1/64 inch diameter, is concentrated. The pressure thus available is about 500,000 pounds per square inch, which is ample to exceed the elastic limit of the hardest and strongest steel in existence.

A remarkable feature of this instrument is that it is self-compensating with regard to the energy of the hammer blows on the softer metals. This is due to the yielding of the material and the comparatively slow stoppage of the hammer. In lead, for example, a deep impression is made. This requires a great amount of energy, which is nearly all spent in doing work, and there is very little rebound afterward — about 3 degrees as against no for the hardest steel. The constant pressure developed by the hammer is thus only 12 pounds instead of 100 or more for good hard steel, and, of course, the pressures for intermediate hardnesses as on brass and soft or tempered steel are always in proportion to the physical hardness of the brass or steel.

Application to Shop Work. — The manufacturer who wishes to get high efficiencies out of his tools will not benefit by the help of such a commodity as the scleroscope in detecting good and bad tools, unless he is willing to amend the errors in practice which he may find. The observation of this principle is the foundation of the success which hundreds of firms in this country and Europe are having with this instrument. While tool work is a line requiring the most careful attention, the material worked and produced is none the less important. In this connection the scleroscope is very commonly applied to industrial systems, with admirable results. An instance may thus be cited showing how these results are obtained.

In 1908, the Brown & Sharpe Mfg. Co. adopted the new method as a guide in the laboratory, particularly for the study and selection of such steel as is required in standard commercial tools. The attention of the company was then turned to its high-grade automobile gears of alloy steels, etc. Meanwhile the Packard Motor Car Co. used the scleroscope to study the past performances of the various gears and parts of old Packard cars, and made careful records. This was also done by many other concerns, and these records showed that alloys, steel or non-ferrous metals would give a certain efficiency if the hardness was just right. As the best is, in the end, the cheapest, in high-grade apparatus, the Packard engineers began to issue orders to their various auto part making houses for material which was specified to require a given degree of scleroscope hardness. Gears were made for them by the Brown & Sharpe Mfg. Co. and the Gleason Works, both of whom are using the scleroscope to aid them in filling orders.

Table 4. Scleroscope Hardness Scale [*]

Name of Metal	Annealed	Hammered
Lead (cast)	2–5	3–7
Babbitt metal	4–9	.....
Gold	5	8½
Silver	6½	20–30
Brass (cast)	7–35	.....
Pure tin (cast)	8	.....
Brass (drawn)	10–15	24–25
Bismuth (cast)	9	.....
Platinum	10	17
Copper (cast)	6	14–20
Zinc (cast)	8	20
Iron, pure	18	25–30
Mild steel, 0.15 per cent carbon	22	30–45
Nickel anode (cast)	31	55
Iron, gray (cast)	30–45	.....
Iron, gray (chilled)	.....	50–90
Steel, tool, 1 per cent carbon	30–35	40–50
Steel, tool, 1.65 per cent carbon	35–40	.....
Vanadium steel	35–45	.....
Chrome-nickel	47	.....
Chrome-nickel (hardened)	.....	60–95
Steel, high-speed (hardened)	.....	70–105
Steel, carbon tool (hardened)	.....	70–105

[*] The figures given are subject to variation, owing to the differences in composition of the metals tested.

Wyman & Gordon, who supply forgings to Brown & Sharpe, were able to make them to the required specifications, but, in order to do so, they had to see that the raw material was of the proper hardness. This brings the matter back to the openhearth or crucible and chemical laboratory, where again the scleroscope is used to great advantage. Before the completion of an automobile of the guaranteed kind, often a dozen instruments are used among the specialty makers who supply the various parts. The ball-bearing manufacturers are required and prefer to test every part before assembling. The Hyatt Roller Bearing Co. and the Hess-Bright Mfg. Co. are obliged to use a number of scleroscopes which are operated by women, carefully trained, who are able to pass on a large number of pieces daily. This testing is to ascertain principally two factors on which success in service depends, *viz.*: the right degree of hardness, and the uniformity of this hardness — and both are equally important. In the latter case it is necessary to test the parts in a number of places, which must be done very rapidly to keep down the additional cost, particularly as in the manufacture of standard parts such as these, there are always losses due to the rejection of some parts which do not conform to the specifications.

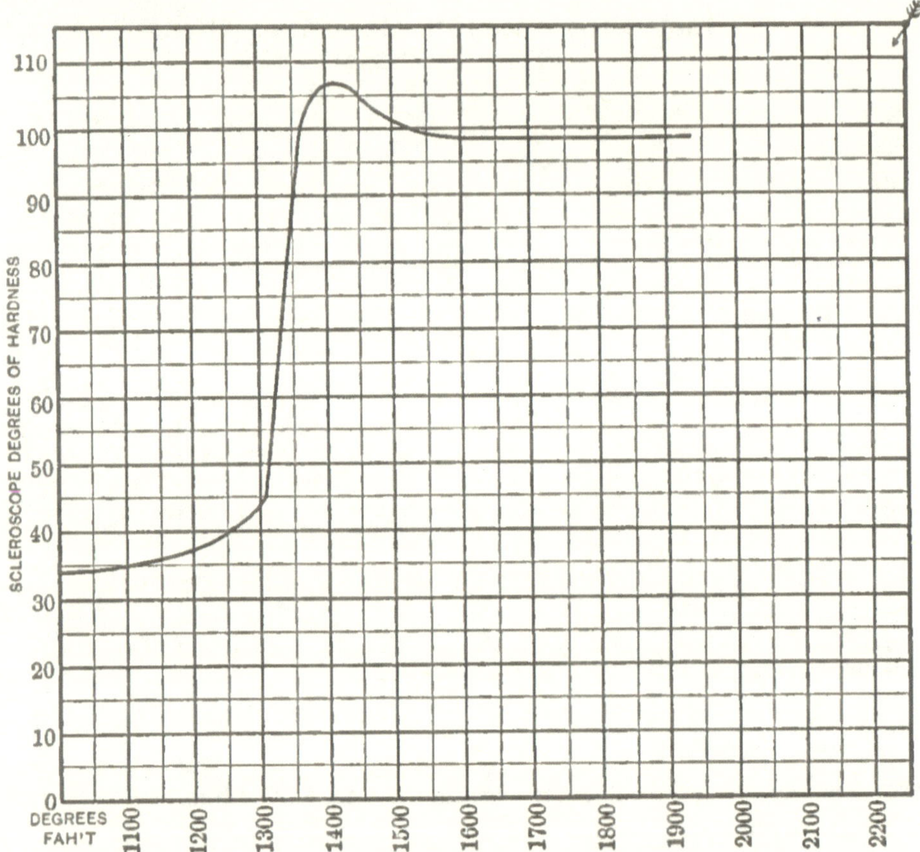

Fig. 9. Hardness Curve for Tool Steel of about 0.90 per cent Carbon

The Lunkenheimer Co., the Light Mfg. & Foundry Co., and other up-to-date manufacturers, use the scleroscope in the standardization of castings adapted to various needs. These houses also make auto parts for the Packard Co., etc., and by the use of the scleroscope are enabled to live up to their specifications. In these auto shops the instrument is used for all classes of work, although it is most needed in the inspection department for the examination of parts and material, particularly of those not made by the builders.

Tool Steel and the Scleroscope. — Since for most uses (other than for turning or planing tools) plain carbon steel is as yet adequate, many manufacturers have turned their attention to the art of obtaining much higher efficiencies by aid of the scleroscope after good steel has been selected. The method of doing this is interesting, and was first hailed as a revelation by many authorities. Thus, when a steel having a carbon content of 0.90 of one per cent and over is heated to the right temperature and is then properly quenched, the limit of hardness and strength is obtained. Now, attaining this temperature is such a delicate matter, that unless the very best facilities are at command, anything but the exact heat required may be obtained, and if the heat is too low the tool will be hard only on the edges, while if it is only a trifle overheated, such as is regularly done by the average hardener who takes chances by depending on the skill of the eye, something like from 50 to 75 per cent of the strength due to rolling or forging is lost. This appalling loss in strength so vitally important in any tool is accompanied by a slight drop in the hardness — not more than 5 per cent.

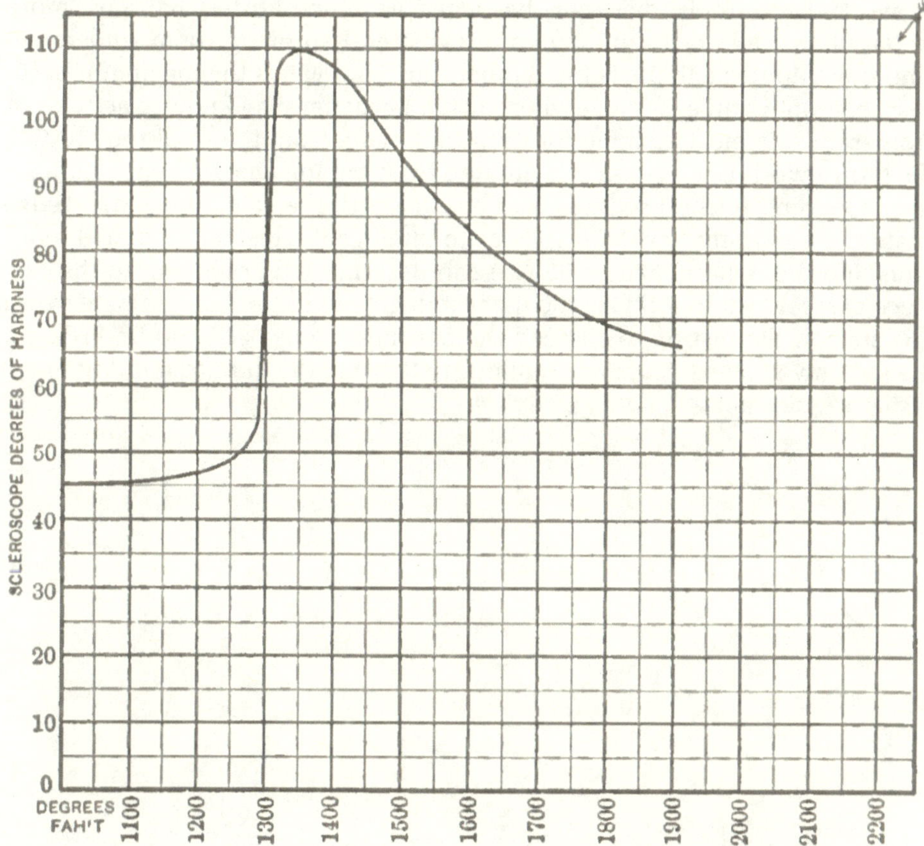

Fig. 10. Hardness Curve for Tool Steel of 1.65 per cent Carbon

This is detected by the scleroscope as shown in charts, Figs. 9 and 10. The former is a hardness curve taken from a tool steel of about 0.90 per cent carbon, while the latter is one taken from a steel having 1.65 per cent carbon. The difference between these curves is indeed very striking and very significant to those who have mastered the elementary principles of the study of tool steels by aid of the scleroscope. The curves are obtained from the Metcalf test as follows: A piece of steel a few inches long and about one-half inch square is heated to a bright yellow on one end and manipulated so that the temperature is less and less toward the other end until a red is scarcely visible. The piece is then quenched in water, ground clean, and tested by the scleroscope at intervals of about 1/8 inch along the bar, beginning at the unhardened end. As each section is tested, a reading is obtained which corresponds with the exact hardness that would be obtained by quenching a similar piece at whatever heat the said test piece had in that location. This hardness number is plotted out on a chart in the usual way so that a true curve is obtained showing the character of the changes in hardness and strength.

The test piece thus obtained represents a stock bar, of the steel which is to be worked into tools, dies, etc., hardened at temperatures that vary more widely than could occur in any well-regulated hardening room, and somewhere within these limits is the temperature that yields the maximum hardness. It supplies an expedient whereby the hardener may know exactly what temperature is most suitable for each tool made from the steel thus tested. His future work is guided and facilitated by stamping on each tool a number corresponding to the hardness number given to the said stock bar. It also enables the hardener or the inspector to intelligently test all hardened tools. Thus, if a die is hardened to 95, we can determine by referring to the test piece of steel, whether this is the highest degree of hardness obtainable with this steel. If the test piece showed the hardness to be, say, 100 or 110, and the die only showed 95, it would indicate that the die did not fulfill the necessary requirements.

www.ingramcontent.com/pod-product-compliance
Lightning Source LLC
LaVergne TN
LVHW090937080826
845145LV00003B/778

* 9 7 8 1 7 8 9 8 7 5 5 6 0 *